Polymer Processing

Polymer Processing

D.H. Morton-Jones

Polymer Research Group, Chemistry Department,
University of Lancaster

CHAPMAN & HALL
London · Glasgow · New York · Tokyo · Melbourne · Madras

Published by Chapman & Hall, 2-6 Boundary Row, London SE1 8HN, UK

Chapman & Hall, 2-6 Boundary Row, London SE1 8HN, UK

Blackie Academic & Professional, Wester Cleddens Road, Bishopbriggs, Glasgow G64 2NZ, UK

Chapman & Hall GmbH, Pappelallee 3, 69469 Weinheim, Germany

Chapman & Hall USA, One Penn Plaza, 41st Floor, New York, NY10119, USA

Chapman & Hall Japan, ITP - Japan, Kyowa Building, 3F, 2-2-1 Hirakawacho, Chiyoda-ku, Tokyo 102, Japan

Chapman & Hall Australia, Thomas Nelson Australia, 102 Dodds Street, South Melbourne, Victoria 3205, Australia

Chapman & Hall India, R. Seshadri, 32 Second Main Road, CIT East, Madras 600 035, India

First edition 1989
Reprinted 1993, 1994, 1995

© 1989 D.H. Morton-Jones

Typeset in 10/12pt Times by EJS Chemical Composition, Bath
Printed in Great Britain by TJ Press (Padstow) Ltd, Cornwall

ISBN 0 412 26700 4

Apart from any fair dealing for the purposes of research or private study, or criticism or review, as permitted under the UK Copyright Designs and Patents Act, 1988, this publication may not be reproduced, stored, or transmitted, in any form or by any means, without the prior permission in writing of the publishers, or in the case of reprographic reproduction only in accordance with the terms of the licences issued by the Copyright Licensing Agency in the UK, or in accordance with the terms of licences issued by the appropriate Reproduction Rights Organization outside the UK. Enquiries concerning reproduction outside the terms stated here should be sent to the publishers at the London address printed on this page.

The publisher makes no representation, express or implied, with regard to the accuracy of the information contained in this book and cannot accept any legal responsibility or liability for any errors or omissions that may be made.

A Catalogue record for this book is available from the British Library

Library of Congress Cataloging-in-Publication Data available

∞ Printed on acid-free text paper, manufactured in accordance with ANSI/NISO Z39.48-1992 and ANSI/NISO Z39.48-1984.

Contents

	Preface	ix
1	**The nature and origins of polymers**	1
1.1	Introduction	1
1.2	Nature of polymers	2
1.3	Basic chemical types	6
1.4	Morphology	12
1.5	Physical properties	17
1.6	Origins of polymers	22
	Reference and Further reading	29
2	**The physical basis of polymer processing**	30
2.1	Introduction	30
2.2	Liquids and viscosity	32
2.3	Viscosity and polymer processing	34
2.4	Other properties of fluids	36
2.5	Shear stresses in polymer systems	37
2.6	Non-Newtonian flow	38
2.7	Practical melt viscosities	40
2.8	Flow in channels	41
2.9	Melt flow index	44
2.10	Melting of polymers	45
2.11	Liquid to solid	47
	References and Further reading	54
3	**Mixing**	55
3.1	Polymers and additives	55
3.2	Physical form of polymer mixes	58
3.3	Types of mixing process	59
3.4	Some processes and machines	61
3.5	Some relationships in mixing	71
	Further reading	73

4 Extrusion — 74
- 4.1 What is extrusion? — 74
- 4.2 Features of a single screw extruder — 74
- 4.3 Flow mechanisms — 82
- 4.4 Analysis of flow — 84
- 4.5 Some aspects of screw design — 92
- 4.6 Twin screw extruders — 97
- 4.7 Extruder and die characteristics — 102
- 4.8 The extrusion die — 105
- References and Further reading — 111

5 Extrusion-based processes — 112
- 5.1 Profile extrusion — 112
- 5.2 Cross-head extrusion — 115
- 5.3 Orientation in pipes and hoses — 117
- 5.4 Orientation and crystallization — 117
- 5.5 Tubular blown film — 118
- 5.6 Other film and sheet processes — 121
- 5.7 Synthetic fibres — 123
- 5.8 Netting — 124
- 5.9 Co-extrusion — 124
- Reference — 125

6 Blow moulding — 126
- 6.1 Blow moulding principles — 126
- 6.2 Extrusion blow moulding — 126
- 6.3 Injection blow moulding — 133
- 6.4 Why PET and why stretch-blow? — 136
- Reference — 137

7 Thermoforming — 138
- 7.1 Principles — 138
- 7.2 Vacuum forming — 138
- 7.3 Material stress and orientation — 142
- 7.4 Applications — 143
- 7.5 Materials — 144
- Reference — 145

8 Injection moulding — 146
- 8.1 Principles — 146
- 8.2 The moulding cycle — 147
- 8.3 The injection moulding machine — 149
- 8.4 The polypropylene hinge–a study in gating — 156

	Contents	vii

8.5	Some aspects of product quality	160
8.6	Sprueless moulding	168
8.7	Newer developments	169
	References	174
	Further reading	175

9 Compression and transfer moulding — 176
9.1	Introduction	176
9.2	Thermosetting compounds	176
9.3	Compression moulding process	179
9.4	Transfer moulding	183
	References	184

10 Polymers in the rubbery state — 185
| 10.1 | The rubbery state | 185 |
| 10.2 | The calendering process | 186 |

11 Rubber technology — 191
11.1	Types of rubber	191
11.2	Production of rubber	194
11.3	Vulcanizing	196
11.4	Fillers	201
11.5	Processing methods	207
11.6	Testing	213
11.7	Thermoplastic elastomers	217
	References	218

12 Fibre reinforced plastics — 220
12.1	Introduction	220
12.2	Materials	220
12.3	Mechanical strength of fibre reinforced composites	223
12.4	The hand lay-up process	226
12.5	Sheet moulding compound (SMC)	227
12.6	Hand lay-up and SMC compared	229
12.7	Dough moulding compound	229
12.8	Process variants	230
12.9	Newer developments using thermosets	230
12.10	Glass mat thermoplastics	233
12.11	Moulding variants	235
	References	236

13 Rotational moulding and sintering — 237
| 13.1 | Evolution | 237 |

13.2	PVC slush moulding	237
13.3	Powdered polymers	237
13.4	Comparison of rotational and injection moulding	240
	Reference	241

14 PVC and plastisols — 242

14.1	Introduction	242
14.2	Polyvinyl chloride (PVC)	242
14.3	Plasticizers	244
14.4	Fillers	245
14.5	Stabilizers	245
14.6	Blowing agents	246
14.7	Substrates	246
14.8	Formulation	247
14.9	Processing	247
14.10	Chemical embossing	251
14.11	The plastigel process	252

Index — 254

Preface

It can be stated with some justification that polymers, because of their mainly synthetic origins, are important because of their applications, perhaps more than in the case of more familiar and conventional materials such as metals and wood, which would exist apart from their use in human activities. The majority of polymers have been synthesized under the impetus of requirements for new and improved properties. The preparative routes to new polymers and blends, and the exploration of their structures and properties constitute absorbing subjects for study, but it is the final application of these materials in real, commercial products that provides the driving force for such developments.

In recent years a number of excellent books have appeared which deal with the chemistry, structure, properties and engineering aspects of polymers. The processing of polymers, as products of the chemical industry, into engineering and consumer goods has received much less attention. There are some valuable texts for individual processes, especially the extrusion and injection moulding of thermoplastics, but others are less well served. This book provided a review of all the important processing routes for transforming polymers into products.

The custom of using the opening chapter of a polymer book to tell the reader what a polymer is has been continued. However, the approach here is rather less chemical than usual. The reader may be an established or student engineer, who will not require detailed chemistry, or may be a chemist, in which case he or she will already possess the necessary knowledge or have ready access to it. More relevant is the characteristic morphology of polymer structures and its influence on their physical behaviour. This has its roots, of course, in the underlying chemistry, and to this extent some chemical understanding is inescapable. Nevertheless, the emphasis is on morphology and the dependence of physical properties upon it.

The second chapter deals with the relationship between processing routes and the rheology of polymers, and also with some aspects of heat transfer. The purpose of the chapter is to relate the behaviour of these materials to their processes in a simple manner which avoids the complexity of more mathematical and specialist texts.

The other twelve chapters are devoted to separate areas of processing technology. The processing methods for thermoplastics, thermosets, rubbers, fibre reinforced plastics and plastisols are described. Throughout, the processes are seen from the standpoint of the processor rather than that of the engineer; although the processing plant is described, the main interest is in how the plastics or rubber material behaves. The book is thus about polymer processing, and not about the construction of the processing plant. This treatment necessitates discussions of some of the materials, because of the way materials and processes interact; however the reader seeking comprehensive treatment of materials should refer to specialized texts.

Although the text is a review of the whole field of polymer processing, the specialist in a particular field should find the account of his own subject a useful summary, perhaps with helpful cross references to related or unfamiliar areas of processing. For example, the treatment of injection moulding in a general work of this length cannot hope to match that to be found in a book devoted entirely to that subject, but the inclusion of the derivative processes such as foam-cored moulding and RIM provides a concise summary of the whole field of injection moulding processes.

The aim has been to provide an outline account of all the major processing routes, for thermosets and rubbers as well as the more familiar thermoplastics. In some, a more quantitative approach has been adopted because the importance of the process has meant that such analysis has been developed, e.g. extrusion, which is seminal to most thermoplastics processes. In other processes a more descriptive account reflects the working practice to be found in the industry.

The book is aimed at undergraduate and postgraduate students who require a general overview of the field of polymer processing in a single easily assimilated volume. It will also be of value to the industrial processor and engineer seeking information about a variety of processes, and as a reference source for the general reader.

I have based this text on the course on processing which I taught as part of the MSc in Polymer Science and Technology (now superseded) at Lancaster University. In this course I was able to draw on the accumulated experience of 30 years' employment in the development of new and improved products in a variety of processing industries, supplemented by the production more recently of a series of intensive case studies of polymer applications. I have tried to pass on something of the endless appeal of these process industries. There is an infinite variety of detail to be tuned for each individual product; no two require exactly the same conditions. Research and Development in these industries requires the matching of the process to the properties of the material, often leading to a modified or even a new process. If this book helps to open the door to this fascinating world it will have succeeded in its purpose.

In conclusion, the acknowledgements: first to my daughter Gillian for her splendid artwork for all the figures, as original drawings; secondly to Mark Kimpton and Barry Statham of Trowbridge College for reading the manuscript and offering helpful suggestions for its improvement, and finally to my wife Jo for her patience through the seemingly endless evenings and weekends when I was hunched over the word processor.

D.H. Morton-Jones
Lancaster
April 1989

1
The nature and origins of polymers

1.1 Introduction

If we compare *polymers* like *polythene* and *nylon* with the more traditional materials used by engineers we find at once several important differences. A few simple tests soon show that polymers

- have lower strengths and stiffnesses
- often have temperature limitations in service – hot or cold
- mechanical tests, e.g. tensile tests, show that they 'creep' i.e., their properties are time-dependent, and this is their most significant mechanical characteristic.

The features listed above are drawbacks in comparison with metals, wood, ceramics, etc. Obviously, polymers must possess advantages in compensation because they are used in vast and increasing quantities, already exceeding and replacing more traditional materials in many key areas: thus by 1981, volume consumption of plastics exceeded that of steel at around 2000×10^3 m^3 p.a., and was increasing while steel consumption declined. What are the advantages offered by these remarkable materials which results in such a rapid expansion in their use? The most important attractions are:

- Polymer materials, both plastics and rubbers, are readily mouldable, which allows easy production of complex shapes with a minimum of fabrication and finishing;
- They have low densities, a property which leads to low weight products;
- They are resistant to corrosion and to chemical attack;
- They are usually electrical and thermal insulators;
- In many applications, the inherent flexibility of polymers is useful – this is especially true for rubbers;
- Although the absolute strength and modulus values of polymers are low, the specific values per unit weight or volume are often favourable. Hence the use made of speciality polymer materials in aerospace applications;
- The special properties of rubbers are elasticity and damping qualities which are applied in springs and energy-absorbing mountings.

2 The nature and origins of polymers

In the early days of plastics and rubbers, the special properties of these materials were ill understood and many early failures occurred because designers did not recognize the significance (or even, often, the existence) of the time-dependent nature of these properties. Nowadays, these properties are well defined and are applied in the design of components to ensure good performance and a long service life. Also, many new types of high-performance polymers have been developed in recent years, which overcome the deficiencies of early types. These are the *engineering polymers*, which offer superior physical properties and enhanced temperature tolerance. Thus, the ready mouldability and insulating and corrosion resistant qualities of polymers can be utilized in more demanding applications.

In this chapter, we shall see how the special and characteristic properties of polymers result from their *chemical* and *morphological* structures; we shall also see the main petrochemical routes in the manufacture of polymers themselves. These then are the raw materials for our main concern, which is the processing of polymer materials into products for engineering or consumer use.

1.2 Nature of polymers

1.2.1 *What is a polymer?*

Essentially, a polymer is a substance whose molecules form long chains, usually several thousand atoms long (Fig. 1.1). The word 'polymer' means

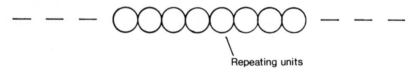

Fig. 1.1 Polymer molecules form long chains.

'many units'. Polymers are characterized, and differ from one another, through the chemical and physical nature of the *repeating units* in the chains. The main interest today is in synthetic polymers such as polyethylene, nylon, synthetic rubbers, melamine laminates, etc. However, there are important naturally-occurring polymers which give a clue to some of the special features of polymeric materials. For example, cellulose is a polymer made by plants in which the long chain-like molecules are linearly aligned to give the plant skeleton linear strength combined with lateral flexibility. This allows the plant to grow tall but to yield without snapping to lateral forces, such as bending in the wind.

Nature of polymers

1.2.2 Some polymer dimensions

What are the actual dimensions of, for example, a polyethylene long-chain molecule of, say, 1000 units (i.e., there are 1000 carbon atoms joined in a continuous chain)? The chemical formula is shown in Fig. 1.2. In fact, this

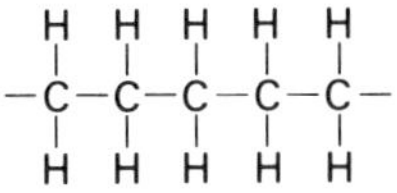

$$-\overset{\overset{H}{|}}{\underset{\underset{H}{|}}{C}}-\overset{\overset{H}{|}}{\underset{\underset{H}{|}}{C}}-\overset{\overset{H}{|}}{\underset{\underset{H}{|}}{C}}-\overset{\overset{H}{|}}{\underset{\underset{H}{|}}{C}}-\overset{\overset{H}{|}}{\underset{\underset{H}{|}}{C}}-$$

Fig. 1.2 Chemical formula for polyethylene.

depiction, though useful to the chemist, does not accurately display the structure of a polyethylene molecule. In reality, the molecule exists, of course, in three dimensions, and the four bonds connecting each carbon to its neighbours are angled symmetrically towards the corners of a tetrahedron, at 109.5°. This can be represented in a three-dimensional drawing (Fig. 1.3).

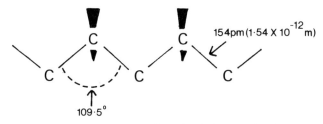

Fig. 1.3 Three-dimensional structure of a polyethylene molecule.

Let us see how large this 1000-unit molecule is. The length of the C—H bonds is 154 pm (1.54×10^{-12} m). If the chain is fully extended linearly, this gives a molecular length of

$$999 \times 154 \times \frac{\sin 109.5}{2}$$

$$= 0.13 \times 10^6 \text{ pm}$$
$$= 0.13 \text{ } \mu\text{m}$$
$$= 1.3 \times 10^{-5} \text{ m}.$$

It is difficult to imagine the physical reality of these dimensions. We know the molecules are elongated, but how can we describe their 'aspect ratio'? A useful analogy is that their L/D (or aspect) ratio is similar to that of a piece of ordinary string about 2 m long. Thus, they are very long and thin. The C—H bond is fixed in length and angle, but it can rotate. The result is that the long, thin chains are flexible. They readily become convoluted and entangled with one another, and as we shall see shortly, their basic physical properties

4   The nature and origins of polymers

derive from this entangled mass. To continue the string analogy, a mass of polymer resembles an entangled ball or web made up of many pieces of string, rather than a box of matches. Such a ball of string would contain pieces of different lengths, and analogously the polymer mass has molecules containing different numbers of units.

1.2.3 *Molar mass of polymers*

In simple materials like salt or phenol or alcohol, all the molecules are the same and have a characteristic *molar mass* (sometimes called by the old name *molecular weight*).

Because a polymer sample contains molecules of different sizes we have to use average values for molecular weight. There is a distribution of molecular sizes and statistical methods are used to express average molar mass. The two most important averages are the *number average molar mass*, M_n and the *weight average molar mass*, M_w.

The number average compares the *total length of polymer chain* with the *number* of individual chains present. The weight average compares the *total length of chain* with the mass fraction of individual chains. Figure 1.4 shows

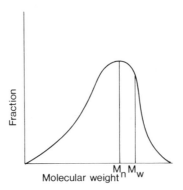

Fig. 1.4 Typical normal distribution of molar mass.

where these two values lie for a polymer with a normal distribution of molar mass. Often, the distribution is asymmetric, and may be broader than shown in Fig. 1.4. Figure 1.5 shows some other common forms of distribution.

The molar mass (or molecular weight, which is the more traditional term) of a particular grade of polymer is an important consideration for processing and performance in service. To define this properly, we need to know not only its value, i.e. whether it is a high or low molecular weight grade, but also the molecular weight spread.

The number average and weight average molar masses can be expressed

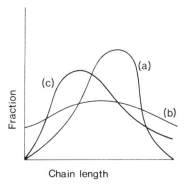

Fig. 1.5 Examples of molar mass distributions: (a) normal distribution; (b) symmetric, broad distribution; (c) asymmetric distribution, a common type.

algebraically as follows. For the *number average molar mass* M_n we find the total mass of polymer chain present and divide this by the number of chains present. Consider an individual chain of length i. Its molar mass is M_i and chains of this size make up a fraction X_i of the total chains present. If we now allow for all values of i, the sum of their masses and their fractions is M_n

$$M_n = \Sigma\, M_i\, X_i.$$

But the fraction X_i is also the number of i-sized chains related to N, the total number of chains. If there are N_i chains of size i

$$X_i = N_i/N$$

and therefore $\quad M_n = \Sigma\, \dfrac{M_i N_i}{N}.$

But $N = \Sigma\, N_i$, i.e. the sum for all values of i, and so

$$M_n = \Sigma\, \dfrac{M_i N_i}{\Sigma N_i}.$$

The number average molar mass thus takes account of the *number* of chains, without regard to their various sizes.

Weight average molar mass, M_w, also uses the total mass (or weight) of the chains, but now, for each individual chain, the fractional *mass*, w_i is used.

$$M_w = \Sigma\, M_i w_i.$$

The ratio M_w/M_n is the *dispersivity*, and is a measure of the molecular weight spread. It is unity only when all the polymer chains are the same length, i.e., the polymer is said to be monodisperse. Normal commercial grades of polymer are never monodisperse, but show a spread of molecular weight which varies according to method of manufacture and finished product and

6 The nature and origins of polymers

processing needs. The dispersivity of so-called 'living polymers', made by anionic polymerization, is often nearly 1, and these polymers are important as speciality products.

1.2.4 Determination of molar mass

Many methods exist for determining one of the average values for molar mass. It is not our purpose to deal with them in detail here, but one which is widely used is outlined below, because it illustrates how simple determinations can be used to find a fundamental polymer property. The method uses *solution viscosity*. When a polymer is dissolved in a solvent, there is a noticeable increase in the viscosity of the solvent. The viscosities of pure solvent and solutions of known concentration can be measured by timing flow through a capillary tube. If the flow time for the solvent is t_o and for a solution of concentration C is t, a number of values can be calculated.

Relative viscosity
$$\eta_r = \frac{\eta_{solution}}{\eta_{solvent}} = \frac{t}{t_o}.$$

Specific viscosity
$$\eta_{sp} = \eta_r - 1 = \frac{t - t_o}{t_o}.$$

Reduced viscosity (viscosity number)
$$\eta_{red} = \frac{\eta_{sp}}{C}.$$

Inherent viscosity (log viscosity number)
$$\eta_{inh} = \frac{\ln \eta_r}{C}.$$

Intrinsic viscosity (limiting viscosity number)
$$[\eta] = \left[\frac{\eta_{sp}}{C}\right]_{C \to 0} = \left[\frac{\ln \eta_r}{C}\right]_{C \to 0}$$

The intrinsic viscosity is found by plotting η_{red}/C against C and extrapolating to zero concentration.

The molecular weight M is then found from the Mark–Houwink equation

$$[\eta] = KM^a.$$

K and a are experimentally determined constants which can be looked up in the literature and M is the *viscosity average* molecular weight, which lies between M_n and M_w.

1.3 Basic chemical types

Many chemical routes exist for the manufacture of polymers, and it is not our purpose here to describe them in detail. Instead the reader is referred to

Basic chemical types

several excellent accounts in the bibliography at the end of the chapter. However, as an introduction for the non-chemical reader especially, the two most widely used manufacturing routes are outlined. These are *addition polymerization* and *step growth* (or *condensation*) *polymerization*.

1.3.1 Addition polymerization

The addition method for making polymers uses as starting material the monomer. As the name implies (one part), this is the small-molecule chemical which will eventually form the polymer by joining up into long molecular chains. The characteristic shared by all small molecules capable of forming polymers in this way is that they are chemically *unsaturated*. This means that they possess a *double bond* between two of their carbon atoms, as shown for ethylene in Fig. 1.6. Although this depiction might suggest that

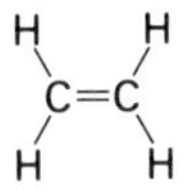

$$\begin{array}{c} H \quad\quad H \\ \diagdown\diagup \\ C=C \\ \diagup\diagdown \\ H \quad\quad H \end{array}$$

Fig. 1.6 Chemical structure of ethylene, showing double bond.

two carbons joined thus are more strongly bonded than a pair joined by a single bond, the opposite is in fact the case. The electronic configuration in a double bond is less favourable and is at a higher energy level than it is in a single bond, with the consequence that a double bond in a molecule is a point vulnerable to chemical attack. This is what happens when an unsaturated monomer is converted to a polymer. The mechanism of the attack involves *homolytic scission* of the double bond to form *free radicals*, i.e., the electron pair comprising the bond divides. The overall effect is for the double bond to be replaced by a single one and the molecule to extend itself, eventually to form a polymer. Figure 1.7 shows the overall result.

$$H_2C=CH_2 \longrightarrow -CH_2-CH_2-$$

Fig. 1.7 Polyethylene forms from ethylene monomer by opening of the double bond.

Ethylene is chemically the simplest of a range of addition polymers. As we have seen, its chemical formula is $CH_2=CH_2$. If one of the hydrogen atoms is replaced by another chemical group a different polymer results. We can use the symbol R to represent the replacement group, and the general formula becomes $CH_2=CHR$. Some examples are given in Table 1.1.

One or two slightly more complex cases may be mentioned here, to show the great versatility of this method for producing polymers. For instance, if the original monomer molecule contains two double bonds, the resulting

8 The nature and origins of polymers

Table 1.1 Some examples of addition polymers

Chemical nature of R	Polymer
Hydrogen, H	Polyethylene
Methyl, CH_3	Polypropylene
Phenyl, C_6H_6	Polystyrene
Chlorine, Cl	Poly vinyl chloride, PVC
Nitrile, CN	Polyacrylonitrile
Methyl acrylate, $COOCH_3$	Poly methyl acrylate
replace a second hydrogen with a methyl, Me	Poly methyl methacrylate

polymer will itself contain double bonds; the polymer is *unsaturated*. As we saw in the basic polymer-forming reaction, double bonds (or sites of unsaturation) are reactive sites in a molecule, which suggests that polymers containing them ought to be chemically reactive, and this is indeed the case. An important example is the range of polymers and copolymers made from *butadiene*, CH_2=CH—CH=CH_2. These comprise the most widely used class of synthetic rubbers. The chemical reactivity of the double bonds is used in the vulcanization of the rubbers with sulphur, when the primary chains are chemically *cross-linked* to one another. Natural rubber has a very similar structure, with double bonds. *Copolymers* use two (or occasionally three) different monomer types in a single polymer, to obtain special properties. An example is the range of ethylene–propylene copolymers in which ethylene and propylene monomers are copolymerized in varying proportions to give different properties in the polymer. A polymer which is nearly all propylene, with about 10% ethylene, behaves mainly like polypropylene; however, its low-temperature performance is improved, at the cost of a small loss in stiffness. When the ethylene content is larger, around 50%, the product is a rubber. Of course, such a rubber cannot readily be vulcanized because it is fully saturated, i.e., it contains no double bonds. To remedy this, a third monomer is used, one of the comparatively rare instances of a *terpolymer*. The third monomer is a diene, i.e., it contains two double bonds. The resultant rubber now has unsaturated sites in it, allowing it to be vulcanized. It is called 'EPDM' (ethylene–propylene–diene–monomer) rubber.

The chains of copolymers may be organized in a number of different ways, which are shown in Fig. 1.8. *Random copolymers* have their comonomer units randomly disposed through the polymer chain. In *block copolymers* the polymerization reaction is controlled to allow many consecutive units to be the same. Thus an ethylene–propylene block copolymer would contain alternating lengths of ethylene units and lengths of propylene units. If the

Basic chemical types 9

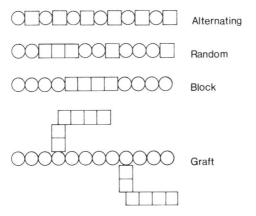

Fig. 1.8 The various types of copolymer.

units alternate singly the result is an *alternating copolymer*. A rather different arrangement occurs when the second monomer type forms a branch chain by attaching at an active site on the main chain: this is a *graft copolymer*.

Copolymers are being increasingly used in sophisticated ways to obtain special properties. The example above is a simple one, in which the low-temperature tolerance of polypropylene is improved. It uses a random copolymer. If, instead, a block copolymerization is used and then, further, we arrange for the small-proportion blocks to be themselves ethylene–propylene random copolymer, the result is a polymer which is mainly polypropylene, but with segments of chain which are EP rubber. This is the basis of the *impact-modified* grades of polypropylene. The impact resistance of these grades is greatly superior, compared with the homopolymer grades, but at the cost of some loss in stiffness.

There are many examples in this active field of development, and the reader is referred to the specialized texts listed in the bibliography.

There is one further important point to make about polymers made by addition. It concerns the spatial arrangement of the chemical groups. As we saw in connection with polymer dimensions, the normal representation on paper of the structure fails to show the three-dimensional arrangement. If we look at the 3 drawings in Fig. 1.9, we see that three variants are possible, depending on the regularity of the disposition of the chemical groups. The most regular arrangement, with all the 'R' groups on one side is the *isotactic* configuration, the arrangement in which the 'R' and H groups alternate is the *syndiotactic* and where the groups are randomly disposed the arrangement is called *atactic*. The older methods for making addition polymers, usually at high temperatures and pressures, led to atactic

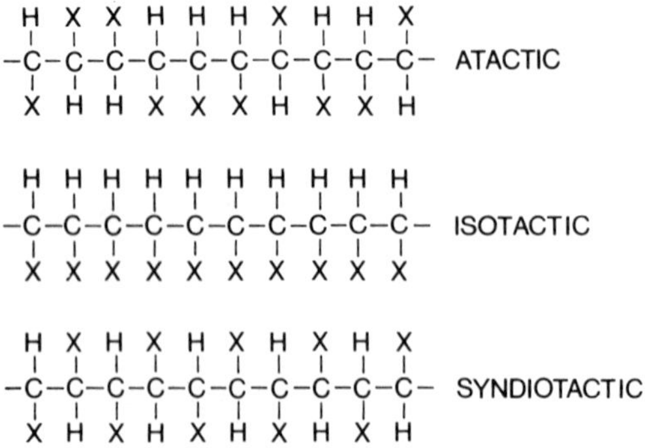

10   The nature and origins of polymers

```
 H X X H H H X H H X
 | | | | | | | | | |
-C-C-C-C-C-C-C-C-C-C-   ATACTIC
 | | | | | | | | | |
 X H H X X X H X X H


 H H H H H H H H H H
 | | | | | | | | | |
-C-C-C-C-C-C-C-C-C-C-   ISOTACTIC
 | | | | | | | | | |
 X X X X X X X X X X


 H X H X H X H X H X
 | | | | | | | | | |
-C-C-C-C-C-C-C-C-C-C-   SYNDIOTACTIC
 | | | | | | | | | |
 X H X H X H X H X H
```

Fig. 1.9 Atactic and stereoregular polymers.

products, but the newer methods using the processes originated by Ziegler and Natta produce *stereoregular* polymers. These processes run at low temperatures and pressures and rely on specially developed *catalysts* for their success. In the case of polyethylene, of course, it does not matter, because R = H, and there is no distinction between regular and irregular structures, which is why the early polyethylene was a successful product. When attempts were made to produce polypropylene by the old high pressure methods the result was the atactic polymer, which is a very unsatisfactory material, soft and without useful mechanical properties. It was only with the advent of the Ziegler–Natta catalysts that isotactic polypropylene became a practical possibility. The reason for the great improvement in physical properties in stereoregular polymers is that the ordered molecular arrangement allows the formation of crystalline regions in the polymer morphology. For the same reason, polyethylene is now made by the Ziegler–Natta process; although stereoregularity cannot be a feature, a nearly fully linear chain results, which is also an important factor in ease of formation of crystalline regions.

1.3.2 *Step growth, or condensation polymerization*

In this route for polymer preparation conventional chemical reactions are used. An example is the formation of esters from acids and alcohols (Fig. 1.10).

$$CH_3COOH + C_2H_5OH \rightarrow CH_3COOC_2H_5 + H_2O$$
   acetic   ethyl   ethyl   water
    acid   alcohol   acetate

Now think of a similar reaction between a pair of bifunctional molecules, e.g., ethylene glycol, which has two alcohol groups, and terephthalic acid,

Basic chemical types 11

$$CH_3COOH + C_2H_5OH \longrightarrow CH_3COO-C_2H_5 + H_2O$$

☐—H HO—◇   ☐—⊢◇ + H_2O

Fig. 1.10 Simple esterification reaction.

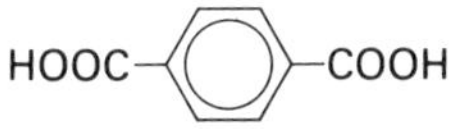

HOOC—◯—COOH   Terephthalic acid

HO—CH_2—CH_2—OH   Ethylene glycol

Fig. 1.11 Bifunctional molecules.

with two acidic groups (Fig. 1.11). The ester produced from these has an alcohol at one end and an acid at the other. Thus it can continue to react to form more ester linkages, with the eventual formation of a long chain polymer molecule. Because the reaction proceeds in this step-like manner this polymer-forming route is called *step-growth* polymerization. The older name for it was *condensation* polymerization, because chemical reactions like esterification are called condensation reactions, characterized by the elimination of a small molecule, in this case water. The preferred modern name for polymer formation by this route, however, is step-growth polymerization. As shown in Fig. 1.12, this type of reaction, between a

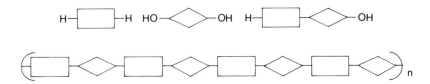

Fig. 1.12 Step-growth polymerization of a polyester.

bifunctional acid and a bifunctional alcohol (called a glycol) leads to the formation of a polyester. The example above gives *polyethylene terephthalate*, *PET*, which is one of the most important examples of this type of polymer. It is widely used as a high strength textile fibre (e.g. *Terylene*, *Trevira*); it is the material for high-quality transparent films (e.g. *Melinex*); in recent years it has also emerged as the polymer for bottles for carbonated drinks and it is now an important material for injection moulded electrical goods because of its outstanding dielectric properties.

Another important polyester is the polymer always referred to simply as 'polycarbonate', but more fully described as the polycarbonate of

12 The nature and origins of polymers

bisphenol A. The latter is a phenol rather than an alcohol, i.e., it is an aromatic compound. It thus contains the benzene rings characteristic of aromatic compounds, and these impart great stiffness to the molecular structure. This contributes to the outstanding impact resistance of this material, although other features of the molecule are also factors. It is also transparent and this combination leads to its selection for applications such as vandal-proof glazing and riot shields. The other major group of polymers made by the step-growth route is the *polyamides*, more commonly known as *nylon*. Here the reacting bifunctional compounds are acids (as in polyesters) and amines, which contain the NH_2 group (Fig. 1.13).

$$HOOC(CH_2)_4COOH + H_2N(CH_2)_6NH_2 \rightarrow \{CO-(CH_2)_4-NHCO-(CH_2)_6-NH\}_n$$

Fig. 1.13 Step-growth formation of nylon 6,6.

1.4 Morphology

1.4.1 *Properties and structure*

In this section we are concerned with the observable physical properties of polymers and their relationship with structure. Section 1.3 gave a brief account of fundamental chemical structure, so let us first see how well this can explain physical properties.

The macromolecular concept for polymers needs to be fitted into a scheme such as that shown in Table 1.2 for a series of molecules of ascending size. In this scheme the properties change in quite a predictable way as we proceed from the gaseous materials with very short carbon chains through the liquid but volatile gasolines, less volatile kerosine, lubricating oils which must be substantially non-volatile to the solid paraffin wax. If this progression were to continue polyethylene would emerge as a substance of almost diamond-like hardness; instead, it is somewhat harder than paraffin wax, but not all that much. The properties are not at all what one would

Table 1.2 The macromolecular concept for polymers

Molecule size	Material	Example of use
C_1–C_4	Gas	Natural and Calor gas
C_5–C_{12}	Gasoline	Petrol engines
C_{15}–C_{20}	Fuels, kerosine	Diesel, heating
C_{20+}	Lubricating oils	
C_{100}	Paraffin wax	Candles
C_{1000+}	Polyethylene	

predict; the material is too soft and too flexible, and its modulus and strength are too low.

If we return for a moment to the 'ball of string' analogy which was useful for envisaging molecular sizes, we at once find a clue to the source of the unexpected properties. The very long, thin chains are convoluted into balls, in an arrangement usually described as the 'random walk'. As with a ball of string, the properties are those of the convoluted mass and not of the individual chains: a loosely and randomly screwed up ball of string is soft, and its qualities have little to do with the ultimate strength of the individual pieces of string. In the polymer, the mechanical response is thus not dependent on fundamental chemical bond strengths, but only on van der Waals forces between chains and their resistance to chain straightening. The van der Waals forces are about 1/100th the magnitude of the primary valence forces which form chemical bonds, and so we have at least a start in explaining the rather unexpected properties of polymers. It does not, however, explain everything as we shall see.

1.4.2 *Amorphous and crystalline polymers*

If we take a list of solid polymers like that shown in Table 1.3 we soon see that the simple explanation above, though a useful start, is inadequate. Another description often applied to the hard, brittle examples like polystyrene and acrylic above is 'glassy', and this is very apposite, because that is exactly what these polymers are. In these materials, the convoluted chain structure is frozen so that molecular motion is greatly restricted. It is as though the pieces of string had become stiff wires. Like all glasses, these polymers are amorphous, i.e., non-crystalline, and also like other glasses, they are brittle and hard. If they are heated they eventually soften and become rather rubbery. This occurs over a quite well defined temperature band, the central point of which is called the *glass transition temperature*, T_g.

Table 1.3 Properties of solid polymers

Polymer	Properties
Polyethylene	Flexible, leathery
Polypropylene	Tough, horny
Polyacetal	Hard, tough, an 'engineering polymer'
Polystyrene	Hard, brittle, transparent
Polycarbonate	Hard, tough, transparent
Acrylic, perspex	Hard, brittle, transparent
Rubber	Highly elastic, low modulus

14 The nature and origins of polymers

Above T_g, the molecular motion is unfrozen and the convoluted chain structure exhibits properties dependent only on the inter-chain van der Waals forces; this is the *rubbery state*. Further heating above the glass transition eventually leads to further softening until the polymer melts. The rubbery region can be seen to form a plateau in a diagram like Fig. 1.14, in

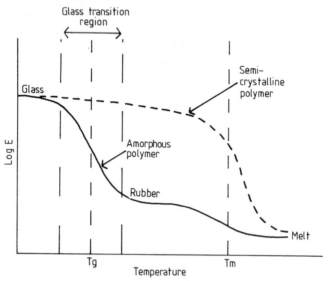

Fig. 1.14 Modulus–temperature diagram for polymers.

which modulus, E, is plotted against temperature. The tough, leathery or horny polymers, like polyethylene and polypropylene, and the harder tough 'engineering' polymers like polyacetal are not brittle at ordinary temperatures, although they may become so at low temperatures. They would thus seem to be above their T_g. Why then are they not soft and rubbery? The reason is that they have appreciable crystalline, ordered regions which impart rigidity to the structure. Such ordered regions are possible when the fundamental chemical structure of the molecule is sufficiently regular. This is why stereoregular polypropylene is a valuable, hard plastic, whereas the atactic polymer is soft and of no practical value. The isotactic or syndiotactic structures allow crystalline regions to be formed because the chains pack together in regular arrays. Similarly, other tough, engineering grade polymers such as acetal, nylon and polyesters, have pronounced crystalline structures. The chemical nature of the nylons and polyesters includes polar groups, which associate to form strong hydrogen bonds between chains and this enhances the strength of the crystalline regions.

These polymers are not wholly crystalline, as for example salt or a simple organic substance might be. The crystalline regions are surrounded by a matrix of less ordered amorphous material. They are above the T_g of the amorphous material, at ambient temperature, and hence derive their hardness from the crystalline regions. Since the amorphous regions are rubbery, to some extent, they may be regarded as composite materials – crystalline and rubber toughened. Alternatively, they can be viewed as rubbery materials reinforced by the strong crystalline regions. Below the T_g of the amorphous polymer they become brittle and glassy.

In a similar way, glassy polymers toughened by added rubbery ones become brittle again below the glass transition of the toughening rubber. An example is high impact polystyrene, HIPS, which is made by adding a small percentage of polybutadiene rubber, usually as a graft copolymer; when cooled below $-80\,°C$ (the T_g of the polybutadiene) HIPS becomes as brittle as normal polystyrene. As one would expect, the hardness of these *semi-crystalline* (as they are more correctly termed) polymers increases with the degree of crystallinity. Their ability to crystallize, as we have already seen, depends largely on the regularity of the basic molecular chains, with polar effects also making an important contribution. The original high pressure polyethylene was a highly branched structure, which would not readily align itself into regular structures, and in consequence has a relatively low level of crystallinity – usually about 20–30%. Another consequence of the branched structure is that the chains are quite widely spaced because the branches interfere sterically; the result of this is a low density, and this material is called *low density polyethylene, LDPE*. It is the polymer used for the traditional polythene bags, rather soft and nearly transparent. The more opaque, tougher bags, with a somewhat crackly feel, are made from *high density polyethylene, HDPE*. This is the polymer made by the Ziegler–Natta route, which it will be recalled gives a nearly linear polymer. The twin results are the higher chain-packing density and a better structure for formation of crystalline regions, with perhaps a 70% level of crystallinity. Similarly, the isotactic polypropylene made by Ziegler–Natta polymerization is about 70% crystalline. However, interestingly, polypropylene has the lowest density of any widely used polymer (0.91), because its molecules contain the bulky methyl side group. Even though the stereoregular structure allows easy crystallization, the crystal unit cell adopts a spacious spiral which gives the polymer its low density.

1.4.3 *Relation of properties to structure*

A number of physical properties of polymers change as the glass transition is passed, in much the same way as they do as the result of a change of state. When a substance melts there is a well defined increase in molecular motion which is marked by phenomena such as latent heat of fusion and changes in observable properties. Thus there are changes in specific heat and the coefficient of expansion. The glass transition similarly is the result of a

16 The nature and origins of polymers

change in the degree of molecular motion, as we have seen, although there is no latent heat and the transition cannot really be regarded as a true thermodynamic change. However, there are observable changes in thermal properties like specific heat and the expansion coefficient. The specific heat change forms the basis of the important technique for characterizing polymers known as *differential scanning calorimetry*, DSC.

An interesting effect of the difference in the thermal expansion coefficient can be seen in a comparison between polypropylene and high density polyethylene. These two materials are close chemical and physical relatives, and indeed they often compete for the same job. It is a matter of frustration and confusion to the designer to learn that their thermal expansion coefficients differ by a factor of nearly 2, and LDPE differs even more from polypropylene. How can this be explained?

Figure 1.15 shows diagramatically how the linear expansion coefficient changes with temperature. Increasing temperature causes an increase in the

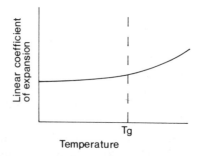

Fig. 1.15 Variation of linear coefficient of expansion with temperature.

expansion coefficient, with a marked change in slope at T_g. In the rubber region the value becomes about 2.5 × the glass value. As a generalization the value, per °C, for most polymers is about 4.5×10^{-5} near T_g and thus becomes $11-12 \times 10^{-5}$ in the rubber region. The value continues to increase at the rate for rubbers (Fig. 1.15). At $T_g + 100\,°C$, the average value is about 22×10^{-5}, and at $T_g + 120\,°C$ it has increased to approximately 25×10^{-5}. Nearly all rubbers are in the region $T_g + 100-120\,°C$ or thereabouts, e.g., nitrile rubber has a value of $19 \times 10^{-5}\,°C^{-1}$. This is useful information for the design engineer, and it is systematic. We can apply it to the polypropylene/polyethylene case cited above.

LDPE is about 25% crystalline, 75% amorphous. Thus it has 75% rubber properties, with T_g about $-100\,°C$. Its expansion coefficient is given by

$$(75\% \text{ of } 25 \times 10^{-5}) + (25\% \text{ of } 4 \times 10^{-5}) \rightarrow 19 \times 10^{-5}$$

This is borne out by observation; LDPE has a coefficient of expansion nearer to a rubber value than a glass one. Now compare polypropylene, which is 75% crystalline, 25% amorphous. Furthermore, the amorphous part has a T_g of $-10\,°C$, which means its coefficient of expansion is only $15 \times 10^{-5}\,°C^{-1}$. This leads to an overall value of 8×10^{-5}, less than half the LDPE figure. In a similar way, HDPE can be shown to have a coefficient of $13–14 \times 10^{-5}\,°C^{-1}$.

1.5 Physical properties

1.5.1 *Morphology and strength*

We have already seen that the dominant influence on polymer behaviour is the morphology, rather than the fundamental molecular structure: the morphology in turn depends, of course, on the chemistry, but the observed properties are those of the coiled and entangled chains and the ways they respond to deforming forces. The most obvious influence of this is found in the strength and stiffness values, both lower by one or two orders of magnitude than strengths deriving from primary bonds, e.g., in metals, ionic solids, carbon fibre. This is still the case even in glassy polymers, which, though stiff, are weak and brittle. The cohesive forces at work are the relatively weak van der Waals forces between chains. Under appropriate conditions, externally applied tensile forces can orientate the chains and straighten them until eventually the primary bond strengths come into play. Such orientation of polymers brings about considerable increase in strength but of course only in the direction of the orientation. Many fabrication processes induce orientation, sometimes giving advantage in a product such as an extruded pipe, but often also creating problems of dimensional instability, especially in injection moulded products.

1.5.2 *Time-dependent properties – creep*

Perhaps the most important and characteristic property of polymers is the time-dependent nature of their physical properties. This becomes very important when a load-bearing product or component is to be designed.

If a component is subjected to a load stresses are created in it and it will deform or deflect, i.e. a strain will result. In most traditional materials such as metals or concrete these quantities are easily manipulated because the materials behave linearly; they obey Hooke's Law, stress $\propto$ strain. Polymers, however, behave differently; they show larger strains and their response is not linear. If the stress is maintained the strain gradually increases – creep occurs. These different types of behaviour are shown in Figs 1.16 and 1.17. A consequence of creep behaviour is that the modulus of

18   The nature and origins of polymers

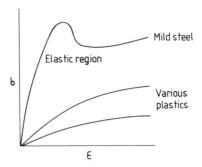

Fig. 1.16 Stress–strain behaviour of traditional and polymeric materials.

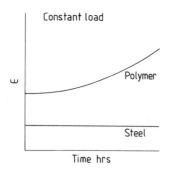

Fig. 1.17 Strain–time curves at constant load.

the material is also not constant but decreases with the duration of application of the load, i.e. stress, $\sigma \propto$ strain, ε

$$\sigma = E\varepsilon$$

where E is Young's modulus. Since ε is not constant with constant σ, E also changes, an effect called *creep modulus*. A further consequence is that if a component is subjected to a constant strain, i.e., ε is kept unchanged, the stress experienced declines. This is known as *stress decay*. These are all modes of the same phenomenon, and they can be represented by a series of diagrams (Fig. 1.18). The *input* quantities are constant with time, and we see how the *output* quantities change. The combined result is the *creep modulus* curve. If this notion is now extended to include a number of different loads a family of curves may be drawn (Fig. 1.19). Curve (b) is obtained from the points on the basic curve (a), using a line at constant strain (isometric). Curve (c) is similarly obtained by a constant time (isochronous) line.

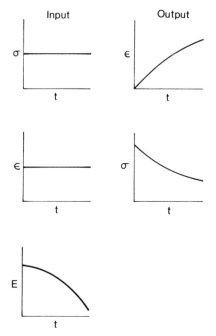

Fig. 1.18 Time-dependency of physical properties.

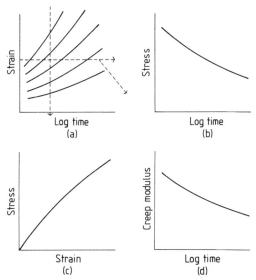

Fig. 1.19 Family of curves showing different representations of creep behaviour in polymers: (a) increasing strains for a number of constant stresses; (b) isometric curve: stress decays with time at constant strain; (c) isochronous curve which reveals non-linear stress–strain response; (d) creep modulus curve which shows modulus decreasing with time.

20   The nature and origins of polymers

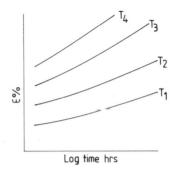

Fig. 1.20 Constant load curves at different temperatures.

These curves show that: (a) a high stress produces a given strain in a shorter time than a low stress; (b) stiffness is greater in the short term than in the long term. This is the creep modulus; the material becomes less stiff the longer it is loaded.

Two further points should be made about creep behaviour. The first is that, of course, different polymers have different families of curves, and the second is that the properties also vary with temperature (Fig. 1.20).

A development of the isometric curve is the *creep rupture curve*. A constant load is applied, straining is continued until failure (rupture) occurs and the time for this is noted. The experiment is repeated with another load and so on for several more loads. The isometric element is the failure strain for each load. From these data a plot can be made of rupture stress against time (Fig. 1.21).

A further variant is the testing of pipes by pressurizing, usually with water, and holding a constant pressure until the pipe bursts. The pressure produces

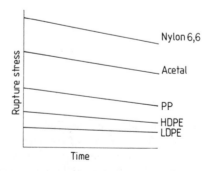

Fig. 1.21 Tensile creep rupture curves for some polymers.

Physical properties 21

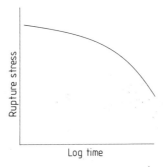

Fig. 1.22 Creep rupture curve for a pressurized pipe at different temperatures.

a hoop stress in the pipe, which is readily calculated from knowledge of the pressure and the pipe dimensions. The experiment may be repeated at different temperatures to give a diagram like that in Fig. 1.22. The curve in Fig. 1.22 represents the results from a series of pressure tests at one temperature. This type of diagram is used in design calculations to find suitable wall thicknesses for extruded pipes. The design life of the pipe is found on the time axis and the design stress can be read off from the y axis: the longer the required life the lower the permitted stress. From this it is a simple matter to calculate the required pipe wall thickness.

$$\text{stress, } \sigma = \frac{Pd}{2s}$$

where P is working pressure in the pipe, d is the pipe diameter, and s is the pipe wall thickness.

The time-dependent nature of polymer mechanical properties is observed because polymers are viscoelastic materials. Viscoelasticity extends across a spectrum (Fig. 1.23). The creep properties we have discussed above are the result of the viscous component, which allows flow to occur. If we return to the morphological model, we can imagine the weakly associated chains tending to yield and slide past one another under a mechanical load. In a fully elastic material the strong primary bonds may stretch under load but they will recover their original configuration when the load is removed. In polymer morphology, there is some elastic response from the greatly entangled chains, but there is also some viscous response from chain slippage. The viscous deformation will be permanent and irrecoverable, but the elastic deformation can recover, sometimes slowly, when the load is removed. Even in glassy polymers some chain slippage can occur under load. At the other end of the spectrum, in polymer melts there is still much

22 The nature and origins of polymers

Fig. 1.23 The viscoelastic spectrum.

chain entanglement, so that polymer melts have quite a pronounced elastic component in their response to loading, e.g., during melt processing. The viscous response of melts is the concern of the next chapter.

1.6 Origins of polymers

1.6.1 *Petrochemistry*

Sedimentary geological deposits 500 000 000 years old form the origins of most of today's synthetic polymers. They were formed from marine organisms, the deposits being washed mainly into continental shelves to form muds in which anaerobic decay occurred. These chemical processes are accompanied by small changes in the chemical free energy, i.e. ΔG is low. The products of such processes are typically carbon dioxide, hydrocarbons, water and ammonia. In silicaceous mud there is about 5% oil and also coalescence can occur. Later earth movements involving impervious clays led to traps of oil and gas in huge chambers in the strata called *anticlynes*, which are domed, and *synclines*, which are concave. These of course have been the main sources of crude oil exploited to date. Where coalescence has not occurred, tar sands are found, and these comprise enormous potential reserves.

Much of the products of the decay is low molecular weight petroleum, which evaporates and recycles instead of being trapped. Also, others of the relatively volatile products of decay evaporate, such as water and ammonia; thus oil deposits contain hardly any oxygen or nitrogen compounds. Compare this with the formation of coal, some 300 000 000 years ago. Coal is formed from cellulosic plant material, which decays to graphitic carbon and methane; there is no evaporation and the reserves are therefore larger.

In the oil traps the expected stable species are found. These are paraffins (chemically saturated hydrocarbons), cyclic paraffins (ring molecules, also saturated) and aromatics (benzene derivatives). There is little material with a high free energy content, like chemicals with double or triple bonds. After the first few carbon atoms in a molecule there is little further increase in free energy and chains containing about 50 carbon atoms can form. Similarly, cyclic compounds with the low-strain configurations of 5 or 6 membered rings are formed, both saturated and aromatic. The composition of the deposits in different parts of the world varies (Table 1.4).

Table 1.4 Composition of oil from oil traps throughout the world

	USA	Near East
C_1–C_4	Much	Moderate
C_5–C_6	Moderate	Much
Aromatics	Appreciable	Little

1.6.2 Exploitation

The earliest small scale, local exploitation was 125 years ago in Canada. However, true commercial exploitation began in Pennsylvania where the oil was virtually seeping out of the ground. The largest known reserves are those in the Middle East. The search for new reserves seems to be undertaken urgently when known resources drop below a critical life of about 30 years. The extraction of the very high quality North Sea deposits was not feasible until the technology became available to combat the daunting problems of extraction in the very hostile marine environment.

A vast and increasing family of industries has developed to use the great array of polymers manufactured from the natural oil base, and sometimes fears are expressed concerning the exhaustion of these reserves. Two important factors operate to render such a catastrophe unlikely. The first is that it is highly probable that there exist large undiscovered new reserves, and the second is that most of the oil extracted is used as fuel. Only a small percentage of it goes down the chemical routes shown in the diagrams on p. 24. Thus, even if the world supply does eventually begin to dry up, so that its use as fuel becomes economically or practically unattractive, the reserves for chemical exploitation are virtually inexhaustable.

1.6.3 Oil distillation

Crude oil is readily distilled to separate it into fractions as shown in Table 1.5. The chemical industry requires a preponderance of the C_1–C_5 fraction, and there is a surfeit of the heavier molecules. The petroleum distillate is therefore cracked into streams, using thermal (700–800 °C), catalytic or fluid bed of Fuller's earth cracking into C_1, C_2, C_3, C_4, C_5 and aromatic streams. These streams are shown in Figs 1.24 to 1.28 [1].

Table 1.5 Fractionation of crude oil

Fraction	Chain length	Use
Gas	C_1–C_4	
Liquid	C_5–C_9	Light naphtha
	C_9–C_{12}	Heavy naphtha
	C_{12}–C_{15}	Aviation fuel, kerosine
	C_{15}–C_{20}	Fuel oil
	C_{20}–C_{35}	Lube oil, waxes
Solid	C_{35+}	Bitumens

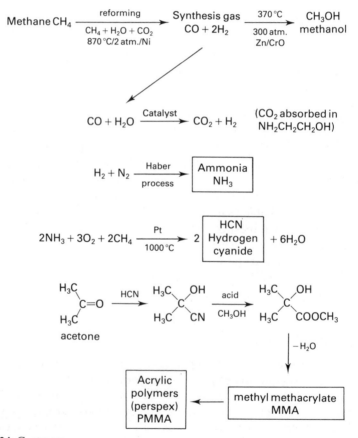

Fig. 1.24 C_1 stream.

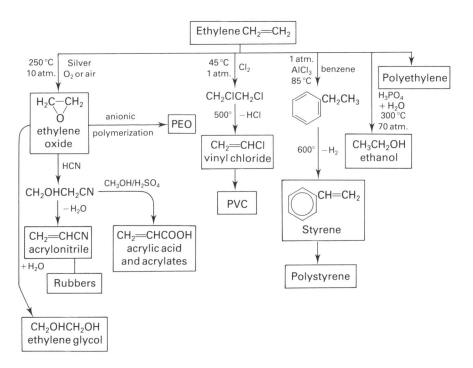

Fig. 1.25 C_2 stream.

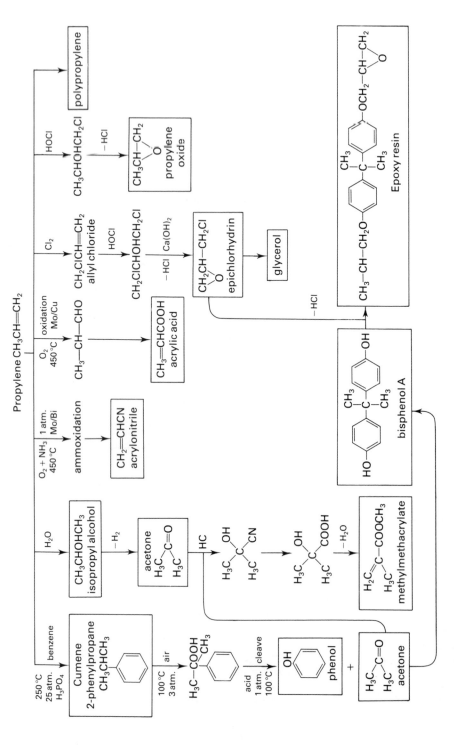

Fig. 1.26 C$_3$ stream.

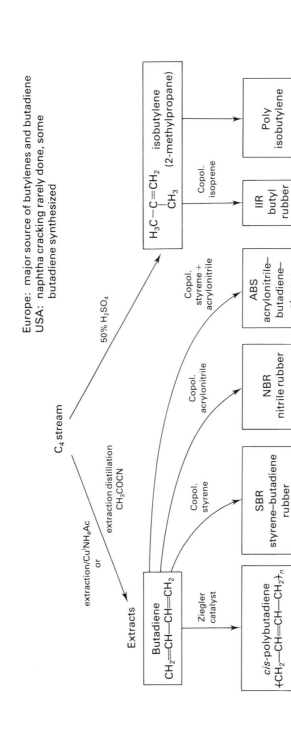

Fig. 1.27 C_4 stream.

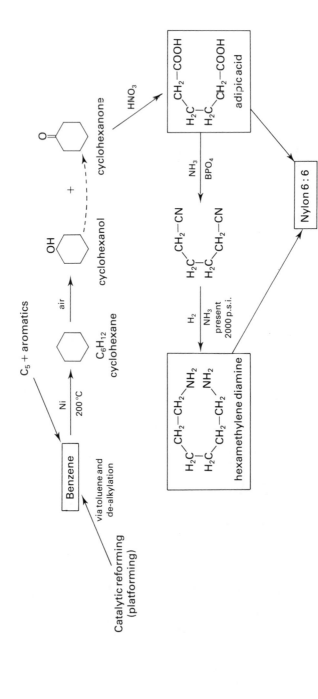

Fig. 1.28 C$_5$ and aromatic streams.

Reference

1. *Organic Chemicals from Petroleum,* BP Educational Service, 1973.

Further reading

Brydson, J.A. (1982) *Plastics Materials.* 4th edn, Iliffe/Butterworths, London.
Young, R.J. (1981) *Introduction to Polymers.* Chapman and Hall, London.
Powell, P.C. (1983) *Engineering with Polymers.* Chapman and Hall, London.
Hall, C. (1981) *Polymer Materials.* Macmillan, London.
Birley, A.W. and Scott, M.J. (1982) *Plastics Materials: Properties and Applications.* Leonard Hill, Glasgow.
Billmeyer Jr., Fred W. (1971) *Textbook of Polymer Science.* 2nd edn, Wiley, Chichester.
Crawford, R.J. (1985) *Plastics and Rubber – Engineering Design and Applications.* Mechanical Engineering Publications Ltd, London.
Nielsen, L.E. (1962) *Mechanical Properties of Polymers.* Van Nostrand Reinhold, New York.
Ward, I.M. (1971) *Mechanical Properties of Solid Polymers.* Wiley, Chichester.

2
The physical basis of polymer processing

2.1 Introduction

As already described in Chapter 1, polymer molecules have long, chain-like conformations which give them their special morphologies. Although chemical in origin, the directly observable and usable properties of these materials derive from their characteristic morphologies. These are the properties which control the processing behaviour and performance in service of polymers.

Let us start a review of processing by looking at a few products and thinking, in outline, how they might be made.

   washing-up bowls
   soft-drink bottles
   rubber tyres
   rubber hot-water bottles
   vacuum cleaner cylinders
   orange-coloured 'Flymo' hoods
   electrical plugs and sockets
   pan and kettle handles
   glass-fibre reinforced tanks and boats

Most people might say, after a little thought, that all these products could be described as having been 'moulded'. Moulding requires the material to be soft and pliable – in fact 'plastic', which is where the generic term 'plastics' comes from. So, it seems, the polymer must have been soft enough, even liquid, to have been mouldable during the vital shaping stage of its processing.

Now look at another list of products.

   paints
   coated fabrics like leathercloths
   cushioned vinyl floorcovering
   glass-fibre repair kits
   hoses
   polythene bags

photographic film
adhesives

Now the 'moulded' criterion is less apparent. However, at least in some of these, common experience reveals a liquid-shaping stage. For example, paints are used as liquids which afterwards harden, and resin/glass-fibre repair kits contain liquid resin which hardens after application. As we shall see shortly, the other products are also made by shaping in soft or liquid form and then solidifying.

If we now look a little more closely at these product lists we shall find in some cases low-viscosity liquids in use during the shaping stage and a profound change, which is irreversible, afterwards in the solidification stage. Examples are

paints
resin/glass products
coated fabrics
vinyl floorcoverings

Rubber products, electrical fittings and pan-handles are similar except that in these cases the liquids are of high viscosity. These are all examples of thermosetting processes, characterized by irreversible chemical change, after shaping, in the solidification stage. In practice, the whole process is carried out in a mould, to give at the end a shaped and solidified product. Usually, the chemical hardening is caused by the formation of chemical cross-links between the polymer chains and it is the function of an additive, such as the 'hardener' in resin/glass systems, or sulphur in rubber, to promote the development of these cross-links.

Turning to other items in the product list we find no liquid precursors. Instead, we find that the products are made by melting the solid polymer. The liquid stage is thus a polymer melt, and this implies processing at quite high temperatures. The solidification stage is then simply to cool to freeze the shaped melt. These are the thermoplastic processes, which are characterized by reversibility of the solid–liquid transition. There are some important consequences of this reversibility.

- Scrap, such as trimmings and defective mouldings, can be re-used by grinding and passing through the process again; this can have an important bearing on the economics of production.
- There is a ceiling temperature for the product in service, which is that at which the material begins to soften again.

The distinction between *thermoplastics* and *thermosets* are summarized in Table 2.1.

When we relate these polymer properties to the processes in use for their conversion to products we find that the processes for the low viscosity

32 The physical basis of polymer processing

Table 2.1 Comparison of thermoplastics and thermosets

Thermoplastics	Thermosets
Use melt in liquid shaping stage	Use lower MW liquid or rubbery polymers at shaping stage
Harden by freezing the melt	Harden by chemical reaction, often cross-linking of chains
Liquid–solid reversible	Liquid goes irreversibly to solid
Scrap recovery possible	Scrap cannot be recovered directly
Ceiling service temperature	Often can withstand high temperatures
Processing of melt usually orientates polymer chains	Can be processed with low orientation

systems like paints and PVC plastisols and resin/glass systems employ simple, low-power plant whereas others, like extrusion and injection of thermoplastics, and rubber processes, need very robust and powerful machinery. To find out why this is so we must look at the rheology of the liquid stages of these various systems.

2.2 Liquids and viscosity

2.2.1 Basic viscosity

In Table 2.2 is a list of substances in order of increasing viscosity. Clearly, the substances become 'thicker' as we descend the Table, but, of course, it is necessary to be more quantitative than this and we shall want to know where the various polymer systems fit. This will begin to give a clue to the magnitude of the forces needed to shape them.

Table 2.2 List of substances: increasing viscosity

Substance	'Thickness' or 'consistency'
Air	Gaseous
Water	Fluid
Olive oil	Liquid
Glycerol	Thick liquid
Golden syrup	Syrupy liquid
Pitch	Flowing solid
Glass	Rigid solid

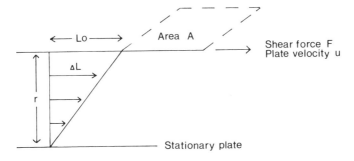

Fig. 2.1 Principle of viscosity.

Consider two plates (Fig. 2.1), separated by distance r. The space between them is occupied by the fluid. One plate moves relatively to the other with velocity U. The movement is resisted by a viscous reaction in the fluid. Since the movement is in shear the reaction is the shear viscosity. We can find the shear stress from the force, F, acting on the moving plate and its area, A.

Shear stress, $\tau = F/A$ Nm^{-2}.

The shear rate is found from the velocity U relative to distance, L: it is thus a derivative quantity

Shear rate, $\dot{\gamma} = U/L$ s^{-1}
$$\tau \propto \dot{\gamma}$$
$$\tau = \eta \dot{\gamma}$$

η, the coefficient of shear viscosity, is the proportionality constant between the shear stress and the shear rate. We now have a quantity which can be used to describe the 'thickness' or viscosity of a fluid.

2.2.2 *Units*

Viscosity is expressed in a bewildering variety of units. Often, it is found by measuring the rate of flow of the fluid or the time of fall of a steel ball through it under standard conditions. Such methods, although valuable for quality and process control purposes, are only comparative and they usually express viscosity in 'seconds', i.e. they simply compare the flow times. For more fundamental study of the rheological properties of materials we need to work in more fundamental units, and the most appropriate are SI units. In the SI system, stress, as expected, will be in units of force per unit area, Nm^{-2}. This is clear from Fig. 2.1, where the shear force, F, acts on area, A.

The shear rate is a derivative; it is the rate of change of velocity of the shearing fluid through the thickness, r. The velocity is maximum immediately adjacent to the moving plate and zero adjacent to the stationary plate.

$$\dot{\gamma} = \frac{dU}{dr}$$

Shear rate thus has the units: velocity/length $= U/L = $ ms^{-1}/m $= $ s^{-1}. This unit is usually called a 'reciprocal second' and shear rates are quoted in reciprocal seconds.

Viscosity is $\tau/\dot{\gamma} = $ Nm^{-2}/s^{-1} $= $ Ns m^{-2}, Newton seconds per square metre. This is often called Pascal seconds, Pas, (1 Nm^{-2} $= $ 1 Pa).

The CGS unit, the Poise, is similarly derived as dyne seconds per square centimetre. The Poise, P, and its derivative the centipoise, cP, are widely used in technical literature. There is, of course, an exact relationship between Poises and Pascal seconds

10 Poise $= $ 1 Pas; cP $= $ mPas (milliPas).

2.3 Viscosity and polymer processing

2.3.1 Shear stresses and shear rates

What are the consequences of viscosity study for the polymer processor? In processing a fluid the 'input' quantity is the shear rate, i.e., the material will be moved, by whatever means is suitable, at an appropriate rate for shaping the article being made. The viscosity may be regarded (for the present) as a material constant property, and a shear stress thus develops in the fluid

$$\tau = \eta \dot{\gamma}.$$

The value of the shear stress clearly depends on the rate of shear and the viscosity of the fluid. If the fluid is of low viscosity, e.g., a paint-like or syrupy liquid, the shear stresses needed to distort it will be small

$$\frac{\tau}{\eta} = \dot{\gamma}$$

i.e., quite low values of τ will generate high values of $\dot{\gamma}$, if η is low.

2.3.2 Shear rates of different processes

Before proceeding with a consideration of the shear rates characteristic of different manufacturing processes for polymer products it is important to remember that high rates of shear are not necessarily involved in high processing speeds, i.e., high volume throughput. The shear rate $\dot{\gamma}$ is concerned with the rate of distortion of the fluid elements.

A few examples of shear rates are displayed in Table 2.3. The shear rate depends principally on the nature of the shaping, fluid-distorting process,

Viscosity and polymer processing 35

Table 2.3 Shear rates of processes

Process	Shear rate (s^{-1})
Compression moulding	1–10
Calendering	10–100
Extrusion	100–1000
Injection moulding	$1000–10^5$
Reverse roll coating	3×10^3

rather than the speed at which that process is run. Much less variation in $\dot{\gamma}$ will result from running the machinery slowly or fast.

2.3.3 *Viscosities of polymer systems*

We can now redraw Table 2.2 to include some polymer systems, and also insert numerical values for viscosity. Table 2.4 is the result. In the table the approximate values quoted for viscosities of polymer systems are those which would be met at processing temperatures. The outstanding feature to emerge from the Table is the huge spread of values, from the water-like latices through syrupy liquid resins to the toffee-like melts and tough rubbers. Clearly, we must expect to find a correspondingly wide variation in

Table 2.4 Viscosities of various materials

Substance or system	Viscosity (MPas)	Consistency
Air	10^{-5}	Gaseous
Water	10^{-3}	Fluid
Polymer latex systems	$10^{-3}–10^{-2}$	Liquid
Olive oil	10^{-1}	Liquid
Paints	$10^{-2}–10^{-1}$	Creamy
PVC plastisols	$1–3 \times 10^{-1}$	Paint-like
Glycerol	10	Thick liquid
Resins for resin/glass	50	Syrup
Golden syrup	10^2	Syrup
Liquid polyurethanes	$10^2–10^3$	Syrup
Polymer melts	$10^2–10^6$	Toffee
Rubbers before cure	$10^2–10^6$	Stiff plasticine
SMC, DMC (moulding compounds)	10^2	Dough
Pitch	10^9	Flowing solid
Glass	10^{21}	Rigid solid

2.4 Other properties of fluids

So far we have only considered the shear viscosity properties of polymer systems. But these materials also have other properties; sometimes they are deformed not in shear but in tension, e.g., in blowing plastic bottles, or making 'blown film' from polyethylene. In these instances the tensile viscosity becomes important. This property is found from the corresponding tensile values of stress and strain rate

$$\text{Tensile viscosity } (\lambda) = \frac{\text{tensile stress } (\sigma)}{\text{tensile strain rate}}.$$

The tensile viscosity, λ, is approximately $3 \times \eta$ for most fluids, including polymer melts. This is one of the reasons why polymer melts cannot be cast by pouring into moulds; as we have seen η is high, and λ is therefore higher, making such a process quite impracticable because pouring involves essentially tensile forces. Much the better approach is to develop processes to exploit shearing forces, i.e. pumping the melt into a mould, which is the principle used in injection moulding.

Also, fluids have definite tensile strength, often about 10^6 Nm^{-2}, compared with around 10^9 Nm^{-2} for solid polymers; the limiting strain is often about 7:1 draw ratio (700%). If these strength properties are exceeded in processing the melt will rupture, and this sometimes occurs. For example, the voids sometimes found in unduly thick-section mouldings are the result of rupture as the moulding cools and contracts from the outside; this imposes tensile forces on the interior, still-molten polymer, which voids when its tensile strength or extension limit is exceeded.

The importance of tensile stress is developed later in connection with particular processes (see especially section 5.5.2).

As well as the viscous response to stressing so far described polymers possess elastic properties, i.e., they are 'viscoelastic' materials. Various models have been used to describe viscoelasticity as exhibited by polymers especially their behaviour in the solid state [1–3]. Although no attempt is made here to give an account of viscoelasticity, it is important to recognize that polymer melts, as well as solids, can exhibit an elastic response when stressed. For example, extruded pipe emerges with a larger diameter than the die through which it was made (die swell). To describe fully the exact response of, for example, a polymer melt including shear viscosity, η tensile viscosity, λ, and shear and tensile elastic moduli, G and E, would be extremely complicated. Fortunately, it is usually clear which of these effects dominates a particular processing situation, and simplified treatments

2.5 Shear stresses in polymer systems

The next step is to see the orders of magnitude of shear stresses which develop in some typical polymer systems. We can find these by considering the shear rates and the viscosities involved. We shall start with low-viscosity systems, like the paint-like plastisols (PVC powder dispersed in liquid plasticizer), the liquid polyester and epoxy resins used in fibre-reinforced articles, liquid polyurethane castings etc.

These processes may be regarded as examples of 'free surface flow' systems. The low-viscosity liquid resins are easily poured or spread by knife-on-roller coating machines on to fabric substrates. Note that both η and λ are low, and pouring is easy.

Assume a viscosity of the order of 3×10^{-1} Pas for, say, a spread plastisol. Such liquids are readily sheared, so shear rates are low, say about 100 s^{-1}. We find, therefore, only low shear stresses are developed

$$\tau = \eta\dot{\gamma} = 3 \times 10^{-1} \times 100 = 30 \text{ Nm}^{-2}.$$

It is quite obvious that relatively light-weight, inexpensive equipment will serve. There are no boundary restrictions, nor are such processes pressure-driven, and the shear stress values give a guide to the forces needed to operate the process.

Now look at a few other typical shear rate/shear stress relationships in processes in Table 2.5. We see a steadily increasing shear stress requirement as (a) the shear rate of the process and (b) the viscosity of the material increase. The situation is further complicated because many processes, such as calendering and injection moulding, involve 'constrained flow'. The material is constrained within boundary walls and is pressure-driven. Further still, in injection moulding particularly, the shear rate can vary over two orders of magnitude during processing, as the melt proceeds along

Table 2.5 Some process shear stresses

Process	Shear rate (s^{-1})	Material viscosity (Pas)	Shear stress (Nm^{-2})
Plastisol spreading	100	3×10^{-1}	30
Plastisol reverse roll coating	3×10^{-3}	3×10^{-1}	900
Rubber calendering	10	5×10^{2}	5000
Injection moulding	10^{3}–10^{5}	150	1.5×10^{5}

38   The physical basis of polymer processing

channels of different dimensions. This, in turn, complicates the viscous response of the melt, and the problems of the process designer.

The shear stresses which develop in materials at higher shear rates and viscosities imply a need for much more robustly constructed machinery. In addition, the driving pressure in, for example, injection moulding leads to a requirement for greater power. We now need to look at the forces required to drive highly viscous materials like polymer melts through narrow channels such as those leading into the mould cavities in injection moulding. However, before we can do this, we must return to the subject of viscosity and introduce 'non-Newtonian flow'.

2.6 Non-Newtonian flow

In the consideration of viscosity given earlier it was assumed that viscosity was a constant property of a fluid. Whilst this is broadly true for many simple liquids, it is not so for more complex materials like polymer melts and dispersions. In many such cases the relationship between shear stress and shear rate is not constant and is not represented as a straight line in a $\tau/\dot{\gamma}$ plot. The linear behaviour, as shown by simple liquids, is known as *Newtonian* behaviour; the non-linear is *non-Newtonian*. There are various classes of non-Newtonian behaviour, depending on whether the material becomes thinner (pseudoplastic or shear thinning) or thicker (dilatant or shear thickening). *Bingham bodies* resist deformation up to critical stress level, after which they yield in Newtonian or non-Newtonian manner. Figure 2.2 shows the various types of response.

The most important type of non-Newtonian fluid is the pseudoplastic one, and most polymer melts and rubber compounds behave in this way. Dilatant behaviour is encountered in some examples of solid-in-liquid dispersions, where it can cause processing difficulties (see Chapter 13).

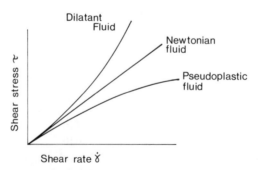

Fig. 2.2 Newtonian and non-Newtonian behaviour.

Non-Newtonian flow   39

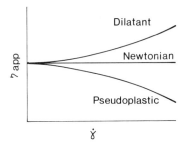

Fig. 2.3 Variation of apparent viscosity with shear rate.

An often more convenient way of displaying non-Newtonian behaviour is to show the variation of viscosity with shear rate (Fig. 2.3). The y axis is labelled η_{app} meaning 'apparent viscosity'. Viscosity is not now a material constant; it has an apparent value for each value of shear rate. We can see from Fig. 2.3 that the apparent viscosity of polymers decreases the more vigorously they are sheared. How can we relate this behaviour to what we already know about the morphology of polymers? As we have seen polymer molecules are long chains, coiled in complex structures. Even in the melt much of this coiling persists and to a large extent this is the source of the elastic component in the response of these materials to stress. However, the more the melt is sheared, the more the coiled chains can be persuaded to slip over one another – the viscosity falls. Just as 'viscosity' can be regarded as the viscous counterpart of elastic 'modulus' so pseudoplastic response is related to the phenomenon of 'creep modulus' observed in solid polymers under a sustained mechanical load, also caused by chain slippage.

A number of complex expressions have been derived to describe accurately the pseudoplastic behaviour of (particularly) polymer melts. Many of these are unwieldy and difficult to apply. In practice, it is found that most polymers can be adequately modelled over a useful range of shear rates by a power law expression

$\tau = k(\dot{\gamma})^n$

where k is often called the *viscosity coefficient* or *consistency index* and n is the *flow behaviour index*. For Newtonian fluids $n = 1$ and k is the shear viscosity η.

The behaviour of a power law fluid may be shown graphically by taking log scales (Fig. 2.4).

$\tau = k(\dot{\gamma})^n$
$\log \tau = \log k \times n \log \dot{\gamma}.$

40   The physical basis of polymer processing

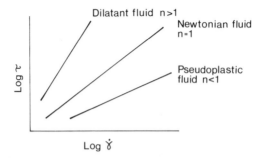

Fig. 2.4 Power law fluids.

Besides varying with shear rate, the apparent viscosity is temperature sensitive. The viscosity coefficient k varies with temperature according to an Arrhenius type equation

$$k_T = k_o \exp(-\alpha \Delta T).$$

2.7 Practical melt viscosities

We have now seen that polymer melts are:

- non-Newtonian
- usually shear-thinning
- decrease in viscosity with increasing temperature.

These features are illustrated in Figs 2.5 and 2.6. The significance of this

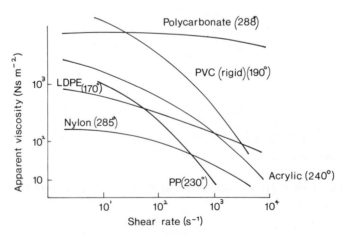

Fig. 2.5 Apparent viscosity vs. shear rate for some polymers.

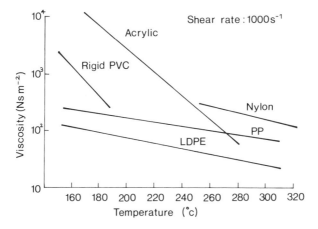

Fig. 2.6 Effect of temperature on polymer viscosity.

appears in high shear rate processes, such as injection moulding and extrusion and many of the processes for rubbers, which are carried out at high temperatures. In estimating the flow behaviour of a polymer in these processes it is no use applying a value for viscosity obtained at a shear rate or temperature quite different from those to be found in the process in question. Curves of the type illustrated in Figs 2.5 and 2.6 may be used to find appropriate values for apparent viscosity. Table 2.6 gives some representative values for the viscosities of a number of thermoplastics at a typical injection shear rate of 10^3 s^{-1}. Table 2.6 reveals very clearly the temperature sensitivity of the viscosity of polymer melts [1].

2.8 Flow in channels

2.8.1 *Cylindrical channel*

We can now return to the subject of flow in the high shear rate processes, using knowledge of non-Newtonian rheology. As an example, consider flow through a cylindrical channel. If length of channel is L, radius of channel is R, pressure drop along channel is P, and volume throughput (per second) is Q, for a cylindrical channel

shear stress at wall, $\quad \tau = \dfrac{PR}{2L}$

shear rate at wall, $\quad \dot{\gamma} = \dfrac{4Q}{\pi R^3}$

Table 2.6 Some representative polymer melt viscosities

Material	Apparent viscosity (Pas) at $\dot{\gamma} = 10^3\,s^{-1}$ and temperature (°C)											
	150	170	190	210	230	250	270	290	310	360	380	400
LD polyethylene	115	85	65	50	40	30	25	20				
PVC	165	100	60									
uPVC		360	310	270								
EVA		220	175	145	115	95						
Polypropylene copolymer			115	105	85	75	65	60				
Polypropylene, 25% glass				145	125	110	95					
Polyacetal			290	200	140							
HIPS				140	115	95						
SAN				210	175	130	90					
ABS				260	195	140						
PMMA				610	300	150	60					
Noryl						240	200	160				
Polycarbonate						790	570	260	190			
Nylon 6,6							115	80	55			
PEEK										480	400	350

(After Barrie [4]).

Flow in channels

apparent viscosity, $\quad \eta_{app} = \dfrac{\tau}{\dot{\gamma}} = \dfrac{PR}{2L} \times \dfrac{\pi R^3}{4Q}$

$$\eta_{app} = \dfrac{\pi PR^4}{8LQ}$$

This is Poiseuille's famous equation for finding viscosity when the dimensions of a channel are known. It forms the basis of capillary viscometry. In this technique, molten polymer is extruded, by means of a piston, through a capillary die. The output rate, Q, is measured, and the die dimensions are known. P is the pressure applied for extrusion. The viscosity may thus be calculated; by varying the pressure it can be found at different shear stresses, which allows the non-Newtonian response to be determined. A simple form of extrusion viscometer is the melt flow indexer which is described in more detail in section 2.9. However first we shall use Poiseuille's equation to find the force necessary to drive a viscous melt through a channel in, for example an injection moulding machine. For the most accurate work, end corrections must be made to account for entry and exit from the channel, but for the present purpose the simple expression will suffice.

2.8.2 To find the driving pressure

If Poiseuille's equation is rearranged to give P

$$\eta_{app} = \dfrac{\pi PR^4}{8LQ}$$

$$P = \dfrac{\eta_{app} 8LQ}{\pi R^4}.$$

Notice the high sensitivity displayed in these expressions to the radius of the channel; R is raised to the fourth power. The shear rate, and hence the apparent viscosity, is profoundly affected by changes in radius, and this will be a critical feature of the design of channels in injection moulds. If the dimensions of the channel are known, and also the apparent viscosity at the appropriate shear rate we can calculate the pressure required to drive the melt through the channel at a required output, Q. This is illustrated in the following example.

A runner leading to the cavity in a mould has the dimensions:
length, $L = 5$ cm (0.05 m);
radius, $R = 0.25$ cm (0.0025 m);
throughput, $Q = 250$ cm^3/s^{-1} (2.50 × 10^{-4} m^3 s^{-1});
viscosity $\eta_{app} = 150$ Pas (typical melt viscosity at $\dot{\gamma} = 10^3$ s^{-1})

$$P = \dfrac{150 \times 8 \times 0.05 \times 2.5 \times 10^{-4}}{P \times 3.9 \times 10^{11}}$$

$= 122$ MNm^{-2} (17 700 p.s.i.).

44   The physical basis of polymer processing

Pressures of this order are commonly used in injection moulding, to inject the melt into the mould runners. As the melt proceeds through the 'gate' into the mould cavity itself, the shear rate increases to about 10^5 s^{-1}. The viscosity, however, decreases, because the melt is shear thinning, and the stress will be of the same order of magnitude.

If Poiseuille's equation is again rearranged to find Q

$$Q = \frac{\pi P R^4}{\eta 8 L}$$

we can readily see how the output rate can be profoundly affected by changes in channel dimensions, especially R, or by changes in viscosity value, say by a temperature change, assuming P to be maintained.

2.9 Melt flow index

A convenient, simple form of extrusion capillary rheometer is the melt flow indexer, which is used to compare grades within a polymer type. In essence, it comprises a cylinder containing polymer melt which is loaded from above by a piston carrying a weight (Fig. 2.7). There is a capillary die at the bottom of the cylinder. The procedure is to drive the melt through the die and to measure the output by cutting off sections of extrudate at known time intervals and weighing them. The die dimensions are known, and the driving weight is known, so that a single-point viscosity could be calculated. However, the melt flow index (MFI) of a polymer is quoted directly as the weight expelled, in grams, in 10 min. In BS 2782, the die dimensions specified are:

length, L = 8.00 mm;
diameter, D (a) (for MFI of 1–25) = 2.095 mm,
(b) (for MFI of 25–250) = 1.180 mm.

The MFI can vary greatly between grades of a given polymer type, e.g., LDPE or HDPE or PP or PS. The principal difference between grades reflected by MFI is in molecular weight, and the MFI can be regarded as a simple index of this quantity, although it does not, of course give any guide to its absolute value. Low MFI means high molecular weight and vice versa. Thus LDPE of MFI = 20 is a low molecular weight grade, whereas MFI = 1.0 indicates a high molecular weight polymer.

The MFI must be treated with caution. It is, after all, only a single point determination, at a not very high shear rate. It cannot be used to compare polymer types; thus MFI values for polyethene cannot be compared with those for polypropylene or polystyrene. However, it is widely used by polymer suppliers to compare grades. Different polymers require different

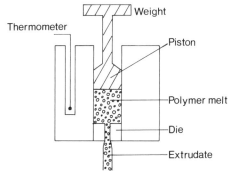

Fig. 2.7 MFI apparatus.

conditions, and standard methods specify a driving weight and a temperature for each polymer type.

2.10 Melting of polymers

2.10.1 *Thermoplastic melting*

Some of the examples considered in earlier sections have been liquid resins or dispersions but others are melts. In the latter case the originally solid polymer must be heated to above its melting or softening point. The heat for this purpose comes from two sources. The first is the external heat supplied, for instance, by electrical resistance heater bands on the barrels of extruders and injection moulding machines, or steam jacketing on internal mixers. The second source is heat generated by viscous dissipation in a highly viscous fluid being sheared at high shear rates. Many internal mixing processes use steam heating only for the first two or three batches of the day and then, once the equipment is fully heated, require only the heat generated by working the polymer. This is particularly the case in rubber mixing – often cooling water is required to avoid too high a temperature and premature vulcanization ('scorch').

The following example illustrates how these two mechanisms for heating operate for a simple polypropylene extrusion. A 1 in. extruder produced sheet through a slit die. The polymer was polypropylene homopolymer, with a specific heat of $1.93 \, \text{kJ kg}^{-1} \, °\text{C}^{-1}$ and latent heat of fusion of $100 \, \text{kJ kg}^{-1}$. The observations listed in Table 2.7 were made. The total energy to heat the polypropylene from 16 °C to 220 °C can be found from its specific heat, latent heat of fusion and the weight produced. The energy derived from viscous dissipation can be found from the power consumption of the extruder when loaded compared with its power consumption empty. The difference is due to the heater bands.

46 The physical basis of polymer processing

Table 2.7 Extrusion of polypropylene homopolymer through a slit die

Observation	A	B
Extruder speed (r.p.m.)	50	80
Current, extruder empty (A)	1	1
Current, extruder loaded (A)	1.5	2.0
Temperature, all zones (°C)	220	220
Ambient temperature (°C)	16	16
Wt of extrudate in 2 min (g)	138	221
Voltage (3-phase supply) (V)	440	440
Power factor of extruder	0.8	0.8

Total heat = [(Specific heat × ΔT) + Latent heat] × weight (kJ).
Viscous heat: power = amps × volts × power factor × 10^{-3} (kW)
energy = power (kW) × time (h) (kW h)
= (kW h) × 3.6×10^3 (kJ).

At 50 r.p.m.
Total heat = [(1.93 × 204) + 100] × 0.138 = 68 kJ

$$\text{Viscous heat} = \frac{1.5 \times 440 \times 0.8 \times 10^{-3} \times 2 \times 3.6 \times 10^3}{60} \text{ (time is 2/60 h).}$$

= 21 kJ.

Viscous heat is 31% of total energy to melt.

At 80 r.p.m.
Total heat = [(1.93 × 204) + 100] × 0.22 = 109 kJ

$$\text{Viscous heat} = \frac{1.5 \times 440 \times 0.8 \times 10^{-3} \times 2 \times 3.6 \times 10^3}{60}$$

= 63 kJ.

Viscous heat is 58% of total energy to melt.

We can see how a considerable proportion of the heat needed to obtain the polymer melt arises from work put into it by the processing equipment. Furthermore, the proportion of heat generated in this way increases as the rate of work increases.

2.10.2 Thermal capacities of polymers

The thermal capacities of polymers are, in general, fairly high, e.g., specific heats, compared with water (1.00) for polystyrene = 0.32, and polyethylene = 0.55. For comparison, that of copper = 0.09.

Table 2.8 Thermal properties of polymers

Polymer	Specific heat $(kJ\,kg^{-1}{}^\circ C^{-1})$	Latent heat of fusion $(kJ\,kg^{-1})$	Process temp. $(^\circ C)$	Total heat to process $(kJ\,kg^{-1})$
ABS	1.47	–	225	300
Acetal copol.	1.47	163	225	465
PMMA	1.47	–	225	300
Nylon 6.6	1.67	130	280	570
Polycarbonate	1.26	–	300	350
Polyethylene, HD	2.30	209	240	720
Polypropylene	1.93	100	250	550
Polystyrene	1.34	–	200	240
uPVC	1.00	–	180	160
Cellulose acetate	1.51	–	195	260
PPO	1.34	–	310	390

Differences between polymers are important in designing processes. As we have seen in the polypropylene example above, the partially crystalline polymers have appreciable latent heats of fusion, which must be supplied during melting in addition to the heat needed simply to heat the mass of polymer. In Table 2.8 thermal properties are given of a selection of polymers. The crystalline ones have values for latent heat, with the result that these materials require greater heating than do the amorphous examples to raise their temperature from ambient to the appropriate processing temperature.

2.11 Liquid to solid

Earlier we saw how the processing of polymers divides into:

- shaping in the liquid (even if very viscous) state,
- solidifying to the final permanent form,

and we have been concerned with the properties of some of these liquid stages, e.g.,

- liquid resins
- plastisol dispersions
- paints
- rubber compounds
- thermoplastic melts.

The next consideration is the various means for solidifying into the final form. Once again, we find two broad divisions. There are the (essentially)

48   The physical basis of polymer processing

liquid systems which solidify by chemical change, and there are the thermoplastics which solidify by freezing.

2.11.1 *Chemical changes in liquid systems*

A wide variety of chemical reactions is employed, and new ones continually appear in the technical literature and in patent applications. A few of the most widely used are listed below. More detailed accounts of the processes themselves will be found in later chapters. For the present, the underlying principles are given in outline. Often, these reactions require elevated temperatures to proceed at a reasonable speed and heated plant is thus frequently to be found, as static or continuous-processing ovens, hot presses, etc. Occasionally, RF heating is used and there is increasing use of radiation curing employing UV or electron beam initiation of radical or ionic species for cross-linking reactions.

Some examples of these chemical stages, often called 'curing' are listed below.

(a) *Liquid resins*

In this case, there is chemical cross-linking. Introduction of additives, immediately before shaping, promotes or accelerates the reaction. *Unsaturated polyester resins* for GRP work use styrene as a thinner, which in addition polymerizes with the unsaturated polyester because it is itself unsaturated. This happens when a free radical initiator, often a peroxide is added. Choice of initiator depends on whether the reaction is to proceed at ambient temperature (hand lay-up, repairs, etc.) or at an elevated temperature (SMC and DMC in hot presses). *Polyurethane* casting and RIM systems depend on mixing two components, just prior to moulding, which react in the mould to give the finished product. *Epoxy resins* are another instance of a two-component system, widely used in mouldings and adhesive.

(b) *Moulding powders*

These are usually based on condensation products of formaldehyde and phenol, urea or melamine and are only incompletely polymerized before being mixed with powdered fillers. Polymerization and cross-linking are completed in the hot mould.

(c) *Plastisols*

These are made from PVC dispersed in liquid plasticizer and they require 'curing' at around 200 °C. At this temperature the PVC fuses and forms a homogeneous matrix with the fully compatible plasticizer. This process is usually carried out by passing the product on a conveyor through a hot air oven. The effect of the plasticizer is to lower the T_g of the PVC; it may be thought of as spacing the polymer chains more widely, with the result that T_g drops below room temperature and the material becomes a rubber. Plasticized PVC differs from the other examples under the present heading because, once fused, it is thermoplastic and not a thermoset. In fact, it is the earliest example of a thermoplastic rubber, and, as such, finds widespread application in, for example, boot and shoe soling.

(d) *Rubbers*

Permanent chemical cross-linking is the characteristic feature of the 'vulcanization' of the conventional rubbers. It is usually conducted at elevated temperatures, under pressure to contain volatile reaction products. The commonest vulcanization technique employs sulphur as the cross-linking agent, with a number of other substances (activators and accelerators) which promote the speed and smoothness of the vulcanization reaction. Variants include peroxide and low-temperature vulcanization for speciality products.

2.11.2 *Freezing of melts*

This is essentially the reverse of the melting process already discussed in outline in Section 2.10. However, there are two important differences from the melting process:

- There is no contribution from work – all the heat must be removed by conduction;
- There is usually no need to return to ambient temperature by the end of the process. Thus, although the moulding or extrudate must be cooled, the target temperature may be higher than ambient. This temperature is usually listed in manufacturers' literature as the 'heat distortion temperature' and it is the maximum temperature at which a moulding may be removed from a mould without danger of its distortion, either during removal or when completing its cooling to ambient by natural cooling.

We now need to examine rather more closely the details of this important feature of plastics processing, using some examples from the field of injection moulding. Similar approaches can, of course, be made for other processes.

2.11.3 Cooling of mouldings

To estimate cooling rates we make use of the quantity *thermal diffusivity*, α

$$\alpha = \frac{K}{\varrho C_p}$$

where K is the coefficient of thermal conductivity, ϱ is the density and C_p is the specific heat.

This quantity may be calculated for any polymer of interest, and values can be found in the literature. However, for practical purposes a value of $10^{-7} \text{m}^2 \text{ s}^{-1}$ may be used for all polymers, as we shall see in the examples below.

Next, the thermal diffusivity is used to find a value for the *Fourier Number*. Fourier's equation for non-steady heat flow in one direction, which is the condition in a cooling moulding, relates dimension and time dependent cooling.

$$\frac{d^2 T}{dx^2} = \frac{1}{\alpha} \left(\frac{dT}{dt} \right)$$

where x is the moulding dimension, T is temperature and t is time.

Solutions to this equation are infinite series and a graphical method is used as follows. Dimensionless groups F_0 and ΔT are found.

$$F_0 = \frac{\alpha t}{x^2}$$

where α is the thermal diffusivity, t is time (of cooling), and x is the moulding dimension (radius of sphere or cylinder, half thickness of sheet).

F_0 is dimensionless ($\text{m}^2/\text{s} \times \text{s} \times 1/\text{m}^2$). It gives an index for the thermal properties, 'time interval' and 'dimension' of a particular situation. When F_0 is small, there is little heat exchange; when it is large (0.8–1), cooling is great. When $F_0 < 0.1$ there is little change; when $F_0 = 1$, the moulding has cooled to surrounding temperature.

The other dimensionless group required is ΔT, which defines the temperature regime.

- How hot is the melt upon entry?
- What is the mould surface temperature?
- What is the demould temperature to be?

From these ΔT is expressed

$$\Delta T = \frac{T_3 - T_2}{T_1 - T_2}$$

where T_1 is the initial melt temperature, T_2 is the temperature of the cooling

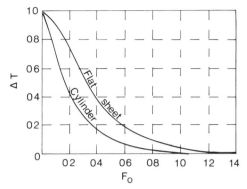

Fig. 2.8 Temperature gradient vs. Fourier number (after Crawford).

medium (mould surface temperature), and T_3 is the temperature of the moulding at time t at dimension x, usually heat distortion temperature. This dimensionless group tends towards 1 when there is no cooling.

The two dimensionless quantities F_o and ΔT can now be represented graphically and used to find cooling times (Fig. 2.8). The following example uses the chart in Fig. 2.8 to explore the cooling times for mouldings under a number of different conditions.

A moulding approximates to a cylinder, 10 mm in diameter. We shall compare its demould times if it is made from (1) polycarbonate, (2) ABS, using appropriate moulding conditions for each polymer.

1. Polycarbonate:

    melt temperature = 310 °C,
    mould temperature = 90 °C,
    heat distortion temperature = 150 °C,
    thermal diffusivity $\alpha = 13.9 \times 10^{-8}$ m^2 s^{-1}.

2. ABS:

    melt temperature = 235 °C,
    mould temperature = 60 °C,
    heat distortion temperature = 85 °C,
    thermal diffusivity $\alpha = 6.5 \times 10^{-8}$ m^2 s^{-1}.

These data will permit comparison of the demould times for the two polymers. By changing some of the parameters, we can also explore the sensitivity of demould time to (a) product thickness, (b) mould temperature, (c) value of thermal diffusivity.

1. Polycarbonate

$$\Delta T = \frac{150 - 90}{310 - 90} = 0.27.$$

52   The physical basis of polymer processing

From Fig. 2.8, cylinder curve, for $\Delta T = 0.27$, $F_0 = 0.32$. Thus

$$0.32 = \frac{\alpha t}{x^2}$$

and $x = 5 \times 10^{-3}$, $\alpha = 13.9 \times 10^{-8}$

$$t = \frac{0.32 \times (5 \times 10^{-3})^2}{13.9 \times 10^{-8}} = 53 \text{ s}.$$

2. ABS

$$\Delta T = \frac{85 - 60}{235 - 60} = 0.14$$

$$F_0 = 0.46$$

$$t = \frac{0.46 \times (5 \times 10^{-3})^2}{6.5 \times 10^{-8}} = 177 \text{ s}.$$

The conclusion from this is that ABS would take roughly three times as long to cool as polycarbonate, using appropriate conditions in the same mould. The next modification is to halve the diameter of the moulding, i.e. $x = 2.5 \times 10^{-3}$.

3. Polycarbonate

$$t = \frac{0.32 \times (2.5 \times 10^{-3})^2}{13.9 \times 10^{-8}} = 14 \text{ s}.$$

4. ABS

$$t = \frac{0.46 \times (2.5 \times 10^{-3})^2}{6.5 \times 10^{-8}} = 55 \text{ s}.$$

Now change the mould temperature for polycarbonate to 60 °C, i.e. the same as for ABS.

5. Polycarbonate

$$\Delta T = \frac{150 - 60}{310 - 60} = 0.30$$

$$F_0 = 0.30$$

$$t = \frac{0.30 \times (5 \times 10^{-3})^2}{13.9 \times 10^{-8}} = 54 \text{ s}.$$

Liquid to solid 53

Finally, assume the 'average for all polymers' value of $10^{-7} \, m^2 \, s^{-1}$ for thermal diffusivity.

6. Polycarbonate

$$t = \frac{0.32 \times (5 \times 10^{-3})^2}{10^{-7}} = 80 \, s.$$

7. ABS

$$t = \frac{0.46 \times (5 \times 10^{-3})^2}{10^{-7}} = 155 \, s.$$

The conclusions to be drawn from these calculations are that demould time is influenced:

- in a major way by the moulding dimensions
- by ΔT, which is mainly dependent on heat distortion temperature
- in only a minor way by mould temperature
- very little by variation in the value of α – the average value will serve for most purposes.

2.11.4 Mould cooling

As we have seen in the examples in the previous section, moulds are held at selected temperatures to optimize product quality and output. How is this achieved? Usually, the mould is supplied with thermostatted water which passes through specially machined channels. An example is given to show the quantities required.

An LDPE moulding consumes polymer at an overall rate of 15 kg h^{-1}. The melt temperature is 190 °C and the mould surface is maintained at 40 ± 2 °C. If the enthalpy of LDPE between 40 °C and 190 °C is 445 kJ kg^{-1}, we can find the water flow rate as follows.

The heat energy to be removed is

$$\frac{15 \times 445}{60 \times 60} = 1.85 \, kW \, (i.e. \, kJ \, s^{-1}).$$

Specific heat of water = 4.186 kJ kg^{-1} °C^{-1}; ΔT allowed for water = 4 °C; therefore enthalpy change, ΔH, for water = 4×4.186 kJ kg^{-1}.
Rate of removal required = 1.85 kJ s^{-1}, therefore

$$\text{mass of water required} = \frac{1.85}{4 \times 4.186} = 0.11 \, kg \, s^{-1} = 0.11 \, litre \, s^{-1}.$$

This moulding process would require cooling water supplied at a rate of 0.11 litre s^{-1} (about 400 litres h^{-1}).

References

1. Young, R.I. (1981) *Introduction to Polymers*. Chapman and Hall, London, Ch. 5.
2. Nielsen, L.E. (1962) *Mechanical Properties of Polymers*. Van Nostrand Reinhold, New York.
3. Ward, I.M. (1971) *Mechanical Properties of Solid Polymers*. Wiley, Chichester.
4. Barrie, I. (1978) in *Polymer Rheology* (ed. R.S. Lenk). Applied Science, Barking. Ch. 13.

Further reading

Brydson, J.A. (1981) *Flow Properties of Polymers*. Godwin, London.
Cogswell, F.N. (1981) *Polymer Melt Rheology*. Godwin, London.

3
Mixing

3.1 Polymers and additives

Before a polymer can be used to make products, it is usually necessary to mix it with added ingredients, which serve a variety of purposes. The mixing processes also provide an opportunity to alter the physical form of the polymer so that it is readily handled at the final conversion stage of its processing, but the primary purpose is the introduction of the additives. Why should it be necessary to do this? There are two answers to this question; the first is that additives are sometimes needed to alter the properties of the material, e.g. by making it harder or more flexible or cheaper; the second is that it is often important to prevent degradation of the polymer in service or during processing or both by means of appropriate additives. These two classifications of additives are listed below.

3.1.1 *Modifying additives*

Modifying additives, as their name suggests, alter the physical properties of the polymer. There are many types and examples of this type of additive include the following.

Reinforcing fillers are used, as their name suggests, to toughen polymers. The outstanding example is carbon black added to rubbers for this purpose. Many rubbers, including natural rubber and styrene–butadiene rubber SBR, the two most widely used examples, are toughened by the addition of quite high proportions of carbon black. The main effect is a marked improvement in abrasion resistance, which is important in many rubber applications, e.g. tyres, conveyor belting. Reinforcement depends on a fine particle size in the added filler, and the other important example is fine particle silica.

Non-reinforcing fillers, or extenders, are also solids, usually powders, added to cheapen the mix or to stiffen it or reduce its tack. These additives do not enhance the properties, in contrast to the reinforcing fillers discussed above. Commonly used materials are calcium carbonate, either as ground limestone or precipitated whiting, and china clay; the latter sometimes has a mildly reinforcing effect if it is fine enough.

Plasticizers are used in cases where it is desired to increase flexibility. They are usually non-volatile liquids. The oustanding example is flexible plasticized PVC; the plasticizer for this purpose is often a high boiling point ester such as dioctyl phthalate (DOP) or tricresyl phosphate (TCP), although others are used for special purposes.

Liquid extenders are often used in rubbers. They are hydrocarbon oils. They extend and cheapen the mix in the same way as the solid extenders, without enhancing properties.

Chemical additives are used to bring about changes in properties. Perhaps the most widely used are those for cross-linking. Here, the polymer chains are chemically bonded to one another at points along their length. The number of such cross-links is decided by the number and distribution of active sites on the polymer chains and the amounts of additives employed. Manipulation of this aspect of formulation is important for controlling the properties of the final product. An example is the 'vulcanization' of rubbers. In this process the rubber chains are cross-linked chemically by sulphur. To do this efficiently a number of other additives are needed (details will be found in Chapter 10). At this stage it is sufficient to observe that this is an obviously sensitive area of formulation. The effect of cross-linking is to increase strength and stiffness, and to reduce creep, because the cross-links impede chain slippage.

Addition of another polymer usually modifies properties, and an important example is the use of small proportions of additive polymers as impact modifiers. These have become widespread in unplasticized PVC (uPVC) in the rapidly developing market for replacement window frames. uPVC has been used widely for many years in buildings, for rainwater goods; its development into window frames demanded new standards of resistance to impact damage and several types of polymeric impact modifier are in use. Other examples of the use of polymer blends are to be found in rubbers, where it is common practice to include two or more types of rubber in the formulation to obtain optimum properties.

Chemical blowing agents are used to produce foamed products. They are substances which decompose chemically at a convenient processing temperature to evolve a gas. The pressure of the evolving gas is balanced by the viscosity of the polymer melt, and a foam is produced. This technique has, of course, been used from time immemorial in cookery for making bread and cakes, where the gas is carbon dioxide from yeast or from the decomposition of baking powder, sodium bicarbonate. In polymer technology, the earliest examples are to be found in sponge, or 'Sorbo' rubber, where the blowing agent is also sodium bicarbonate. For higher temperature melts like the thermoplastics other blowing agents have been developed, usually evolving nitrogen. The method has been developed into a number of very sophisticated areas (see chapter 8 on injection moulding variants and the 'chemical embossing' or plastisols in chapter 14).

A final example of modifying additives is the use of pigments or dyes to colour the product.

3.1.2 Protective additives

There is a very large number of additives in this classification. Some representative categories are as follows.

Antioxidants are used to protect against atmospheric oxidation. Many polymers have sites on their molecular chains which are susceptible to attack by oxygen. An example is the 'tertiary hydrogen' atom (marked with an asterisk) in polypropylene (Fig. 3.1). The electronic configuration of this

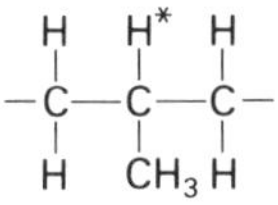

$$-\overset{\overset{\displaystyle H}{|}}{\underset{\underset{\displaystyle H}{|}}{C}}-\overset{\overset{\displaystyle H^*}{|}}{\underset{\underset{\displaystyle CH_3}{|}}{C}}-\overset{\overset{\displaystyle H}{|}}{\underset{\underset{\displaystyle H}{|}}{C}}-$$

Fig. 3.1 The vulnerable tertiary hydrogen in polypropylene.

site leads to ready loss of this hydrogen, and then oxidation occurs. A series of reactions follows, resulting finally in chain scission. Each time scission occurs the chain is shortened, with consequential degradation of properties. It is the function of antioxidant additives to combat oxidative attack, by interfering chemically with the series of reactions which lead to scission. Antioxidants are needed to prevent oxidation during processing and also to protect the polymer structure during the service life of the product. During processing polymers are often exposed to quite severe regimes of temperature and shear, allowing ready attack by oxygen. Without protection, it would be quite impossible to process polypropylene satisfactorily, for example, because the loss in properties would be unacceptable.

Heat stabilizers perform a similar function in preventing degradation at high processing temperatures. Their function is to stop other types of degradation reaction, for example the tendency of some polymers to depolymerize, or 'unzip'. These additives are particularly important in PVC, which readily degrades and darkens when heated, with the evolution of hydrogen chloride. A comprehensive range of very effective heat stabilizers has been developed to combat this.

Antiozonants are a type of specialized antioxidant used especially in rubbers. The unsaturated double bonds in rubber molecules are very susceptible to attack by ozone, even in the low concentrations found in the air. The attack becomes especially severe when the rubber is stressed. The

58   Mixing

effect may be seen on rubber articles exposed to the air out of doors as small surface cracks. Without antiozonants this type of attack can lead to rapid failure of rubber goods.

UV stabilizers often work in conjunction with antioxidants. One of the initiating reactions, starting an oxidative degradation, is attack at the reactive site on the polymer by UV radiation, in sunlight. Thus, this becomes important in exposed outdoor situations, especially where there is a lot of sunlight. UV stabilizers work by absorbing the UV preferentially, and re-emitting the energy harmlessly at a lower wavelength.

Antistatic agents are sometimes helpful in preventing the build-up of undesirable static charges, either during processing or in service. Such charges can make processing difficult and their discharge can be unpleasant. They can even create a potentially dangerous spark in some cases, e.g. where flammable solvents are in use.

Processing lubricants are widely used to assist the passage of the material through the processing machinery. They are often oils or waxes, and a huge range of proprietary brands exists to deal with widely differing product and processing conditions. They are usually classified into two groups.

1. Internal lubricants, which lubricate the polymer granules, and those of other additives, during processing. This allows easier and cooler melting, with a reduced risk of thermal damage. These materials are often at least partially miscible with the polymer melt.
2. External lubricants are essentially immiscible. They lubricate the mix against the processing machinery, allowing the correct degree of friction for the process to work but preventing too much friction which will also cause local high temperature and degradation. It is quite important to get the proportion of lubricant right in a formulation because too much impairs good mixing and subsequent processing.

3.2 Physical form of polymer mixes

The type of plant required for the production of mixes of polymers and their additives depends on their physical form. Raw polymers are supplied in a variety of forms which include large (25 kg) bales of solid rubber, granules, powders of various particle size characteristics, liquid medium molecular weight resins which extend or cross-link during subsequent processing, latex (plural: latices), which is a colloidal dispersion of solid high molecular weight polymer in water. Natural rubber latex, from the rubber tree *Hevea brasiliensis* is a naturally occurring example which is familiar in concentrated form as an adhesive like *Copydex* or *Revertex*. Clearly, the machinery required to break down tough bales of rubber and then to mix in vulcanizing ingredients, reinforcing fillers, processing oils, various antioxidants and stabilizers without overheating will differ from that for adding pigments, thickeners, UV absorbers, etc. to a water-based latex to make a simple latex paint.

In a few cases, the physical nature of the mix depends more on the additive than the polymer; an example is the blend of powdered PVC and liquid plasticizer, termed a *plastisol*, which assumes the properties of the plasticizer and has a paint-like consistency.

3.3 Types of mixing process

Before proceeding to discussion of individual mixing processes, we must determine what it is they are required to achieve. The means for dealing with the variants outlined in the preceding sections obviously must involve very diverse individual processes. However, we can identify two basic mixing functions, and the individual processes set out to accomplish one or other of these for their materials and conditions. Each of the two functions, rather confusingly, carries a number of names:-

- extensive mixing
- blending
- mixing
- distributive mixing

- intensive mixing
- compounding
- dispersion
- dispersive mixing

3.3.1 *Extensive or distributive mixing or blending*

Distributive mixing, also known by the other names listed, consists essentially of stirring together the ingredients. Often this will mean the blending of a number of solids, e.g. polypropylene powder, pigment, antioxidant. The result is a mixture of powders; the individual powders remain and can, in principle, be separated, although in practice this might be difficult. Inspection under a magnifying glass will reveal the individual powder particles. A large sample, in terms of numbers of particles, is needed to give a statistically representative composition of the blend. Small proportions of liquids can be added; they are usually adsorbed by the solid ingredients.

3.3.2 *Intensive or dispersive mixing or compounding*

Dispersive mixing involves the more intimate dispersion of the additives into the matrix of the polymer. It usually requires:

- a physical change in the components
- high shear forces to bring about the change
- the polymer to be in the molten or rubbery state during mixing.

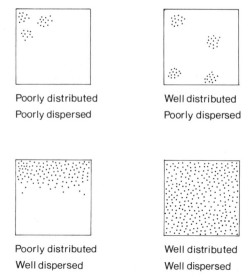

Fig. 3.2 Diagrammatic representation of distributive and dispersive mixing.

Suppose the resulting 'compound' is afterwards granulated; each granule will now have the same composition, assuming the mixing to have been well done. In practice this may be something of an ideal – the requirement may well be to mix at proportions of 1 in 1000. Nevertheless, this is the aim, and where additives have to act together it is important to get near to perfect dispersion, and distribution. The distinction between distribution and dispersion are shown diagrammatically in Fig. 3.2.

3.3.3 When are 'blending' and 'compounding' used?

In most cases it is necessary to achieve both good distribution and good dispersion for a satisfactory product.

Blending or distributive mixing is used:

- When the fabricating process to follow offers some compounding action. For example, pigments may be 'tumbled' into granules or powder prior to extrusion or injection moulding to give a coloured product.
- Thermosetting powders are often blends of powdered resin and fillers which disperse upon fusion of the resin during moulding.
- As a preliminary to a separate compounding process.

Compounding or dispersive mixing is needed as a separate process:

- When accurate distribution and dispersion of interactive ingredients is required. An example is found in rubber compounding where four or five

additives have to act together to give smooth and efficient cross-linking of the rubber. They must be evenly distributed and finely dispersed through the matrix to achieve this.
- When large amounts of modifying ingredients, e.g., fillers, plasticizers, other polymers, etc., are being used.
- When the fabricating process offers little or insufficient compounding action.

3.4 Some processes and machines

The above sections have classified the aims of mixing and some of its basic variations. We can now turn to a few of the myriad processes used to achieve these aims.

3.4.1 *Blending*

Processes for blending vary from the simplest to sophisticated high speed machines.

The simplest is to tumble together dry ingredients, for example, in a 40 gallon drum on a pair of rollers. This is rather a slow process but it can be effective if rapid throughputs are not required and the demands are not too critical. It can be improved by the use of a twin-drum tumbler (Fig. 3.3).

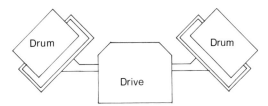

Fig. 3.3 Twin-drum tumbler.

In the *ribbon-blender*, a tumbling action similarly takes place. However, the chamber is stationary and the ribbons rotate, constantly scooping the material from the outside to the centre (Fig. 3.4).

These blenders can be jacketed for steam or electrical heating, when they can be used for producing PVC 'dry blends'. The latter contain powdered PVC polymer, plasticizer and various other ingredients such as stabilizers in smaller proportions. The PVC must be the correct grade (suspension-polymerized, easy-processing grade). When these ingredients are blended at about 100 °C the plasticizer is adsorbed by the polymer and a free-flowing powder results which is ideal for feeding to, e.g., extruders, internal

62   Mixing

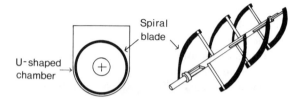

Interrupted spiral blade

Fig. 3.4 Ribbon blender.

compounding mixers (see below), or melt coating machines. The ribbons in these machines move relatively slowly, (e.g. 100 r.p.m.). They are cheap, simple, and easy to run.

A more sophisticated and rapid machine for blending, and especially widely used for PVC dry blends is the *high-speed mixer*, originated by Henschel and often referred to as a *Henschel mixer*, but also made by other manufacturers such as Fielder and Papenmeier. A typical design is shown in Fig. 3.5. These machines run at several thousand r.p.m. and form a

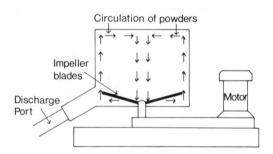

Fig. 3.5 High-speed mixer.

circulating vortex of powder which becomes heated by friction, up to 150–200 °C in many cases. This makes them particularly suitable for making PVC dry blends because external heating is unnecessary. They can be water jacketed for running cold with heat-sensitive polymers. High-speed mixers are widely used for PVC dry blends, drying, incorporating pigments, antioxidants, etc. in preblend powder. They are frequently used as preliminaries to compounders, described below.

Dough-like blends, for example *dough moulding compound, DMC,* are frequently made in *Z-blade mixers* (Fig. 3.6). DMC is a blend of syrupy

Fig. 3.6 Z-blade mixer.

unsaturated polyester resin and fillers, especially short-staple glass. Such a blend cannot be satisfactorily produced in a ribbon blender or a Henschel-type mixer because of its doughy consistency. In the Z-blade mixer the two Z-shaped blades counter-rotate to distribute the solid fillers into the liquid base. These mixers were originally developed for use in the food processing industry and they are particularly effective for making blends of doughy consistency. Notice that these mixes are still blends. No physical change has occurred in the polymeric dispersion medium nor in the additives.

Paints and the paint-like plastisols are prepared in *paddle-mixers*, which are like scaled-up domestic food mixers; food mixers are often used for laboratory-scale mixing. The viscosity of such systems is low enough for these relatively low-powered machines to be effective. When the solid ingredients of such a paint require further refining a *ball mill* may be used. This device comprises a cylindrical vessel containing a large number of steel or ceramic balls in a range of sizes (Fig. 3.7). As it rotates about its axis the

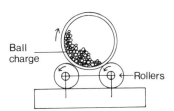

Fig. 3.7 Ball mill.

balls tumble inside, together with the paint or powder. Agglomerates of powder are broken down by the grinding action of the tumbling balls. The resulting refined paint is very smooth, although the process is slow, several hours milling usually being needed. Ball milling may be regarded as a simple compounding process since the physical nature of the added solids is modified by it.

64   Mixing

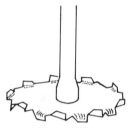

Fig. 3.8 Cowles dissolver.

Low viscosity liquid blends are made with a *dip mixer*. Typical is the Cowles dissolver (Fig. 3.8). In these devices the stirrer is dipped into the mix contained in a 'change can'. This denotes that the container may be changed and the mixer is a simple dip-in stirrer. The characteristic of a dip-mixer is that it is a low shear device. This makes it particularly suitable for latex or emulsified mixes, which are often used as surface coatings; such colloidal dispersions are easily coagulated by high shear mixing.

The machines described above illustrate the main classes of process for distributive or extensive mixing. There are many detailed variants from as many suppliers, often designed to meet special problems, e.g. especially difficult viscosity conditions or special temperature sensitivity. It is impossible in a general review to list all of these. The reader seeking more detailed information should turn to the technical literature from individual machinery suppliers.

3.4.2 Compounding – rubbers

By contrast, compounding or intensive mixing employs high shear processes and much more powerful machinery. Heat is often required to obtain a polymer melt – although this is not always the case, as we shall see in the case of rubber compounding.

The simplest, and basic, machine for intensive mixing is the *two-roll mill* (Fig. 3.9). The two-roll mill is comprised of a pair of rollers with axes horizontally disposed to one another, giving a vertical 'nip' between them.

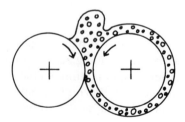

Fig. 3.9 Two-roll mill.

The polymer matrix and additives are subjected to high shear in the nip as the rolls rotate in opposite directions. The rolls may rotate at the same speed, or sometimes with a speed ratio between them which may variously be 1/1.1 to 1/1.4. The resultant mix is an intimate one, below the resolution of the eye. Two-roll mills do this well in the machine direction but poorly across (i.e. along the rolls). Thus we can say that this machine is good at intensive or dispersive mixing but poor at extensive or distributive mixing; it produces a well dispersed but poorly distributed mix, unless the processor takes steps to correct matters. How this can be done will become clear as we describe the use of a two-roll mill for mixing.

When a two-roll mill is used for mixing, the technique is to pass the appropriate loading of matrix material, usually raw polymer, through the nip a few times until it warms up, softens and forms a smooth band round one of the rolls. Often the rolls are preheated, by steam or electrically heated oil; the temperature required depends on the individual polymer properties. Which roll forms the band depends on the polymer and the conditions, but in general it is the hotter, faster roll.

The bearings of the rolls are held in movable bushes which can be used to adjust the nip gap. The nip is adjusted, once the band is formed round the preferred roll, to give a small 'bank' of polymer rolling along the top of the nip. As soon as this condition is achieved, the additives can be introduced. This is done by distributing them manually along the length of the nip. The mill immediately begins to incorporate them into the material on the mill, as 'band' and 'bank' material interchange in the rolling bank. This process is assisted manually by cutting the band with a knife from one edge to two-thirds to three-quarters of its width, so that a flap of it is formed which can be folded to the other side. This allows rapid exchange of bank and band material and also improves lateral distribution of the batch. By cutting and folding many times, from both sides, good distribution and dispersion are achieved. The entire batch can be removed from the mill by cutting across the full width and pulling off the batch as a continuous sheet; if the batch size is too large for this to be convenient, it can be removed by cutting suitably sized sheets until the whole batch is off.

Two-roll mill mixing started in rubber processing. Mills of many different sizes exist for various functions. The biggest production machines have 84" rolls, other standard sizes being 60", 48", and 36". Smaller mills are used in pilot-scale and laboratory work, where 18", 12" and 6" are common sizes.

As might be expected from the description of the process, mixing on the two-roll mill is time consuming – 2 h for a 200 kg mix on a 84" wide mill – and it depends for its success on considerable skill on the part of the mill operator. It is no longer in use as a primary production process, although still widely used for laboratory scale work, and sometimes for one-off trial batches.

66 Mixing

The mill is still widely used as the receiver of material mixed in *internal mixers*. In this role it is important:

- as a refiner of the mix
- as a cooler – the internal mixer produces large masses of hot mix which often require cooling, especially rubber compounds
- as a convenient way of turning the large chunks into easily handled sheets
- for the addition of sensitive ingredients; for example, vulcanizing additives for rubbers are often added on the mill to reduce the chance of premature reaction ('scorch').

The advent of the internal mixer revolutionized the scene, and the original type, the *Banbury mixer* still dominates the picture. In outline, the Banbury mixer has the following features (Fig. 3.10).

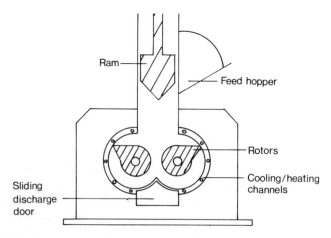

Fig. 3.10 Banbury mixer – diagrammatic.

- There are two rotors, counter-rotating within a chamber;
- Each has two or four 'blades' which mix by smearing the material against the chamber wall;
- A weighted ram keeps the mix in place inside the chamber;
- Uniformity of mix is achieved by a swirling action caused by the speed differential of the rotors;
- There is a complex flow pattern within the mixer, with elongational flow as material enters the nip between the rotors and shearing flow as it leaves it;
- The rotors and the chamber walls are cored for steam or water for heating or cooling.

Banbury mixers are of very stout construction because the forces

Table 3.1 Approximate capacities of mixing plants

Machine	Mill size (in.)	Capacity (kg)
No. 11 Banbury	84	350
No. 9 Banbury	84 or 60	200
No. 3 Banbury	36	80

generated during mixing a tough rubber compound are considerable. They are manufactured in a range of sizes, still often denoted by a numbering system which was originally devised to indicate the number of 60″ two-roll mills the mixer would replace. This system is now obsolete and the suppliers of these machines (Farrel-Bridge in the UK) have adopted a more logical system which specifies volume capacity. As a guide, the approximate capacities of typical mixing plants are shown in Table 3.1 for a highly filled rubber stock.

More recently, an alternative design of internal mixer has been pioneered, especially by Francis Shaw Ltd., whose model is called the Intermix. This has much the same general appearance as the Banbury, but has the following variants:

- Massive rotors with flat topped projections ('nogs') which intermesh;
- Mixing is mill-like, between the rotors;
- As in the Banbury, a ram operates to keep the material between the rotors;
- The resultant thin layers are alleged to allow better cooling, (important for rubbers) by virtue of a larger cooling surface;
- This is claimed to allow higher rates of energy input before a critical temperature is reached.

What do these internal mixers offer the compounder, compared with his simple two-roll mill?

First and foremost they vastly increase the rate of throughput, and, properly run, the regularity of the product. We have already seen that a 200 kg batch of a highly filled rubber compound would take 2 h to mix on an open two-roll mill. A number 11 Banbury mixer will produce 350 kg in 15 min or less. Of course, the same amount of energy is required to mix a given batch, and internal mixers are consequently, as we have seen, of very stout construction and require high power inputs; a number 11 Banbury will be powered by a 500 or 1000 hp motor, depending on the toughness of the mix being produced.

As one would expect, a characteristic of mixing under high shear conditions is heating by viscous dissipation of energy. This heat is useful in softening the polymer prior to the introduction of fillers and other additives.

68 Mixing

In fact for some polymers, e.g. PVC, it is augmented by steam heating of the mixing chamber and rotors to give the correct conditions for mixing.

For rubbers, however, it is usually necessary to limit the temperature rise resulting from viscous heating:

- to avoid the risk of premature cross-linking, the rubber processor's nightmare, known as 'scorch'
- to maintain good high shear mixing; if the rubber becomes too soft, the shear stress goes down and the dispersive mixing action is reduced.

For these reasons, mixers for rubber are more usually found operating with water cooling than under heat.

The processes considered so far are primarily used for rubbery polymers. These include the rubbers themselves – natural and synthetics, and also PVC compounds for calendering into sheets and extrusion into profiled lengths, e.g. window frames, rainwater goods, hoses, pipes, cable coverings, sheeting for apparel and industrial use. Most of these rubber and PVC compounds contain major modifying additives:

- the vulcanizing system for rubbers
- plasticizers in PVC
- fillers – reinforcing and non-reinforcing
- other polymers

and many more. Compounds require constant monitoring and detailed change. An inherent feature of product design in these fields is compound formulation to meet product specification. Thus these processes are conducted 'in house' and the mixing and compounding departments, often termed 'primary processing', are very important in such industries.

A typical *mixing procedure*, using an internal mixer is outlined below.

- The polymer is added to the empty mixer first, to warm up or to break down the 'nerve' of a rubbery polymer.
- If more than one polymer is in use, they are blended at this stage until a homogeneous blend is achieved.
- Additives are added gradually to maintain coherence of the polymer phase. Thus additives to be used in small proportions would be added first to ensure good distribution and dispersion through the polymer matrix. Bulky fillers would then be added in two stages. This is to make sure that the continuous polymer phase is maintained. Once the coherence of the polymer phase is lost, it is usually impossible to recover it. The result is a 'crumbed' mix, usually to be scrapped.
- Various means are employed to judge the end point of a mix. In some instances it is possible simply to time the additions and the total duration of mixing. Other techniques are to observe the current being used by the motor, which rises and falls as mixing proceeds, or to observe the

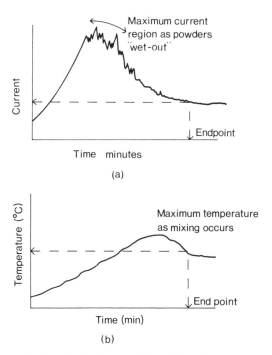

Fig. 3.11 Possible end-points for Banbury mixing: (a) using current consumption; (b) using mix temperature.

temperature of the mix itself and to 'dump' the mix at an appropriate point. Figure 3.11 illustrates some typical patterns.

3.4.3 Compounding – thermoplastics

For the typical moulding and extrusion grades of thermoplastic polymers a rather different situation holds. Included here are, for example: polyethylene, polypropylene, polystyrene, polycarbonate, nylon, acetal, acrylics, and a host of others. In these fields, the final processor making, e.g., extruded profiles, injection mouldings, blow-moulded bottles, thermoformed baths, is not at all concerned with the preparation of the polymer for its final process and product. There are usually no major modifying ingredients. The processor selects a suitable grade of polymer for his purpose and simply processes it.

Now, of course, as we have seen, the polymer will need to contain its protective additives such as antioxidants, UV stabilizers, colorants, lubricants, but these are required in standardized and much smaller proportions compared with the situation for rubbers. Even the recent emergence of many glass, talc (and others) filled grades of thermoplastics

70 Mixing

only involves standardized proportions, and the final processor does not mix in these additives himself. In general, the typical injection moulding, extrusion or blow-moulding factory will have no compounding facilities at all, and no expertise in this area. The onus for producing and supplying polymers containing appropriate additives usually falls upon the polymer manufacturer; there are also some specialist compounding companies.

The user will usually require his selected grade to be supplied in easily usable 'nibs' or granules, which are suitable for the feed hoppers of extruders and injection moulding machines. The easiest way to achieve all this is to use an *extruder* as a mixer. Often, the extruder is itself fed with a blend from a Henschel-type high speed mixer. Before we deal with the use of the extruder for mixing, it is important to understand how an extruder works. In fact, as we shall see, the simple, single-screw extruder is a rather inefficient mixer and numerous developments and modifications have been introduced over the years to improve its effectiveness. An account of mixing with extruders will therefore be found in the next chapter, on extruders and extrusion.

3.4.4 *Mixing processes – summary*

Figure 3.12 summarizes in diagrammatic form the various routes for mixing and compounding discussed in more detail above.

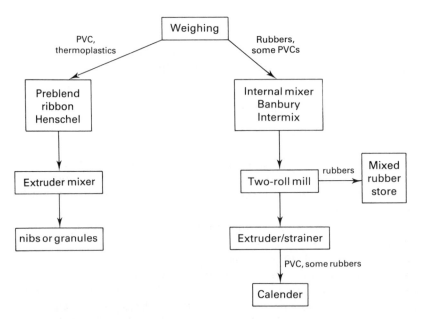

Fig. 3.12 Mixing schemes for different classes of material.

3.5 Some relationships in mixing

Before leaving this subject, it is of interest to explore the quantitative relationships which describe the behaviour of these materials during mixing processes.

3.5.1 Forces in mixing

The first question is 'how is the force transmitted, to break down agglomerates of additive particles?' The answer is 'by fluid mechanical stress in the mixer'.

Consider two spherical additive particles, radii r_1 and r_2

$$\text{stress } \tau = \frac{\text{force}}{\text{area}} = \frac{F}{a}$$

a (for the two agglomerate particles) $= 3\pi(r_1 r_2)$, and thus

$$F = 3\pi\tau(r_1 r_2)$$

Since

$$\tau = \eta\dot{\gamma}$$

$$F = 3\pi\eta\dot{\gamma}(r_1 r_2)$$

Energy dissipated per unit volume, $P = \eta(\dot{\gamma})^2$

and thus

$$P = \frac{F^2}{9\pi^2 r_1^2 r_2^2 \eta}. \tag{A}$$

The term for viscosity η is in the denominator, hence less energy is needed under high viscosity conditions to achieve good dispersion. This is another reason for keeping rubber mixes cool, besides the need to avoid scorch; the higher viscosity under cooler conditions allows more efficient mixing.

3.5.2 Routes for mixing

If we return to the diagrammatic representation of mixing in Fig. 3.2, and expand the concept a little, we can think of two distinct routes; which will be appropriate in a given situation will depend on the viscosity regime of the system. This is shown in Fig. 3.13, which is derived from Fig. 3.2. Route 1, with well distributed but poorly dispersed additive, will entail lower viscosity than Route 2. Equation A suggests that Route 1 will require more energy than Route 2. When the dispersive stage offers *relatively* low viscosity, e.g. a polymer melt in an extruder, the distributive function to break down the

72 Mixing

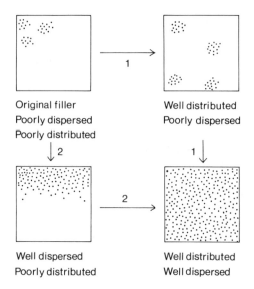

Fig. 3.13 Routes to mixing.

agglomerates is poor and a *preblending* stage is needed. In this context, a high viscosity is that of a rubber, and the single stage Route 2 will work.

A familiar everyday example is found in bakery, the baker preferring to knead a stiff dough rather than to beat a thin batter to break down agglomerates of flour.

But, we may ask, why is distributive mixing so difficult to achieve in melts? What emerges is a *threefold* hierarchy of viscosity regimes. *Low viscosity systems*, e.g. latices, low molecular weight prepolymers, plastisols, paints etc., permit *turbulent* mixing. The boundary between *laminar flow* and *turbulent flow* is described by Reynold's equation

$$Re = \frac{DV\varrho}{\eta}$$

where V is the velocity of the fluid of density ϱ and viscosity η down a circular channel of diameter D.

Re, the *Reynolds Number*, which is dimensionless, must exceed about 2000 to achieve turbulence. This is easily attained in low viscosity systems with normal stirring speeds. However, consider the example of a channel where:

$D = 0.5$ cm $= 0.005$ m,
$\eta = 150$ Pas,
$\varrho = 1000$ kg m^{-3},
$Q = 250$ cm^3 s^{-1} (Q is the volume throughput).

First we must derive the velocity from the volume throughput.
Since $D = 0.5$ cm, A, the cross-sectional area of the channel,

$$\pi r^2 = \pi \times (0.0025)^2$$
$$= 1.96 \times 10^{-5} \text{ m}^2.$$

$$V = \frac{Q}{A} = \frac{2.5 \times 10^{-4}}{1.96 \times 10^{-5}} = 12.7 \text{ m s}^{-1}$$

$$Re = \frac{0.005 \times 12.7 \times 1000}{150} = 0.42.$$

Now, the internal mixer is not a simple channel; nevertheless, it is clear from this low value of Re that turbulent flow cannot occur at flow rates likely to be generated in polymer melts. Thus, distribution via turbulence cannot be a feature of mixing in these systems. As we have already seen (Equation A), at even higher viscosities the input energy is used efficiently to bring about both distribution and dispersion in a single process. Our three viscosity regimes are thus seen to be:

- At very low viscosity, turbulence results in efficient distribution.
- At high viscosity, as found in polymer melts, turbulence cannot occur and distribution is poor, although dispersion is quite efficient.
- At very high viscosity, as in rubbers, there is sufficient shear to break down agglomerates, and efficient distribution and dispersion can occur in a single process. Even so, as we saw in the account of two-roll mill mixing, such processes do not inherently bring about good distribution, and the requirements of distribution and dispersion are to some extent in conflict.

Further reading

Matthews, G. (1982) *Polymer Mixing Technology*. Applied Science, London.

4

Extrusion

4.1 What is extrusion?

In principle, the extrusion process comprises the forcing of a plastic or molten material through a shaped die by means of pressure. The process has been used for many years for metals such as aluminium which flow plastically under deforming pressure, and in the earliest form of extrusion process for polymers similar ram-driven machines were used. In the modern process, however, screws are used to progress the polymer in the molten or rubbery state along the barrel of the machine. The most widely used type is the single screw machine and Fig. 4.1 shows the main features of this type of extruder. Twin screw extruders are also used where superior mixing or conveying is important, and they will be discussed in Section 4.6. The machine consists essentially of an Archimedian screw fitting closely in a cylindrical barrel, with just sufficient clearance to allow its rotation. Solid polymer is fed in at one end and the profiled molten extrudate emerges from the other. Inside, the polymer melts and homogenizes. What is the action of the screw and how is the material conveyed by it to force it through the constricted die?

4.2 Features of a single screw extruder

The screw of an extruder has one or two 'flights' spiralling along its length. The diameter to the outside of the flight is constant along the length to allow the close fit in the barrel. The 'root' or core, however, is of varying diameter and so the spiralling channel varies in depth. In general, the channel depth decreases from feed end to die end (Fig. 4.2) although there are variants for special purposes. A consequence of the decreasing channel depth is increasing pressure along the extruder (Fig. 4.2) and this is what drives the melt through the die. There are three zones, whose functions are as follows. There is also the die zone, which will be examined separately.

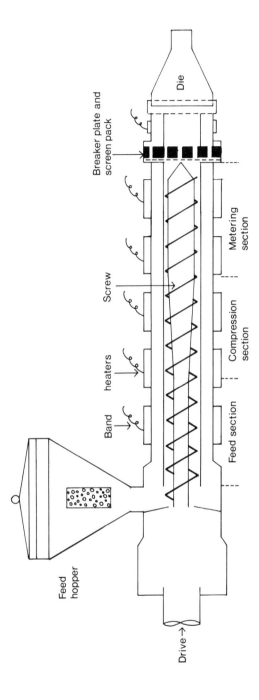

Fig. 4.1 Main features of a single screw extruder.

76 Extrusion

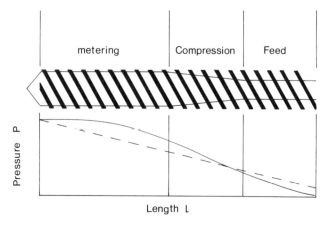

Fig. 4.2 Zones in a single screw extruder.

4.2.1 *The zones in an extruder*

(a) *Feed zone*
In the first zone, usually termed the 'feed' zone, the function is to preheat the polymer and convey it to the subsequent zones. The screw depth is constant and the length of this zone is such as to ensure a correct rate of feed forward, neither starving it nor overfeeding. This varies somewhat for optimum performance with different polymers.

(b) *Compression zone*
The second zone has decreasing channel depth. There are several functions for this zone, usually called the 'compression' or 'transition' zone. Firstly, it expels air trapped between the original granules; secondly, heat transfer from the heated barrel walls is improved as the material thickness decreases; thirdly, the density change during melting is accommodated. Again, there is variation in the ideal design for each polymer type. For a polymer which melts gradually, e.g. low density polyethylene, a screw like that shown in Fig. 4.2 with the overall length roughly evenly divided between the three zones is appropriate. Screws of this type are often referred to as *polyethylene screws*. If the polymer melts relatively sharply the conventional wisdom is that a very short compression zone is needed, usually only one turn of the screw flight in length; an example of such a polymer is nylon, hence the common name *nylon screw* for this design. However, there is little theoretical justification for this view and these polymers perform well in

Features of a single screw extruder

continuously compressing screws [1]. Nevertheless, rapid compression screws are widely used for nylon and other semi-crystalline polymers such as polypropylene and acetal. PVC is a difficult polymer to extrude, melting even more gradually than polyethylene. It is really a thermoplastic rubber and has unusual frictional properties; it is often best processed using a screw which is one long compression zone along its entire length, sometimes with the addition of a metering zone. These variations are shown in Fig. 4.3.

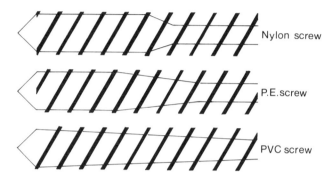

Fig. 4.3 Variations in screw design.

(c) *Metering zone*
Once again we find constant screw depth. The function is to homogenize the melt and hence to supply to the die region material which is of homogeneous quality at constant temperature and pressure.

4.2.2 *The die zone*

The final zone of an extruder is the *die zone*, which terminates in the die itself. Located in this region is the *screen pack* (Fig. 4.1). This usually comprises a perforated steel plate called the *breaker plate* and a sieve pack of two or three layers of wire gauze on its upstream (screw) side.

The breaker plate–screen pack assembly has three functions:

1. To sieve out extraneous material, e.g. ungelled polymer, dirt, foreign bodies;
2. To allow head pressure to develop by providing a resistance for the pumping action of the metering zone;
3. To remove 'turning memory' from the melt.

78   Extrusion

Looking at these a little more closely:

1. Sieving helps to reduce downsteam product faults by removing unwanted particles. It is surprising how often metal particles or even small nuts or bolts will be trapped on the screen, as well as, for example, agglomerates of filler that have escaped dispersion. Apart from causing product faults, metallic inclusions will damage the die, which is usually a serious matter because dies are expensive and difficult to repair.
2. The importance of developing a head pressure is that it is this pressure that provides the driving force for the die.
3. In many cases, the polymer will 'remember' its history of turning along the spiral screw, even after it has been through the die, and this can result in a twisting distortion in the product. Polymers, as we have seen, are made up of long chain molecules, coiled and intertwined even in the melt; this is the source of their viscoelastic behaviour. The melts, though predominantly viscous, have significant elastic properties as well. When the melt receives a prolonged mechanical treatment, such as a passage down a screw, appreciable alignment of chains occurs; this manifests itself as a tendency towards elastic recovery of this alignment as the preferred energetic configuration. Passage through the die is relatively short-lived, without time to replace the spiralled configuration by a new one. The result is a tendency for the product to twist once it escapes the constraint of the die and before it hardens.

4.2.3 *An example of 'turning memory'*

A practical example of this phenomenon is described below.

Some years ago, the author was involved in the development of a new design of flooring block. It was to be a form of parquet flooring made in coloured, highly-filled PVC compound. The manufacturing process was to compound the PVC, plasticizer, calcium carbonate filler, heat stabilizers and pigments and then to extrude a strip of the required profile. As the strip emerged from the die, it would be cut by a pair of knives to give rectangular blocks, which could then be cooled in a water bath. Figure 4.4 shows the product and process diagrammatically. The blocks or tiles had a cross-sectional profile suitable for laying them in a cement slurry and keying them permanently to the sub-floor.

During the early pilot scale trials, a rather old extruder was pressed into service and a die was made. At this early stage there was no breaker plate! Once the preliminary trials to establish running temperatures, speeds, etc. were complete and a continuous extruded strip was obtained, the first tiles were cut. It was quickly realized, when they were removed from the cooling bath, that they were all twisted. When placed on a flat surface they would

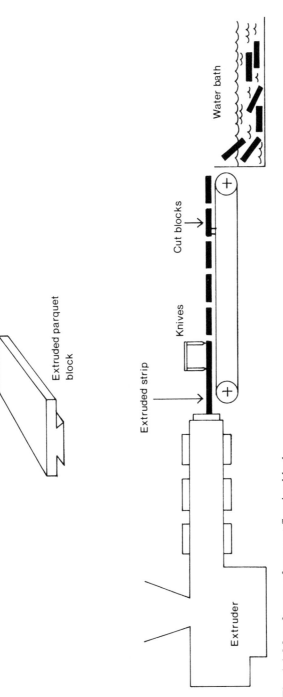

Fig. 4.4 Manufacture of parquet flooring blocks.

rock slightly! Such a defect was, of course, fatal because it would not be possible to produce a good level floor with them.

Closer inspection of the twisting revealed that it was always in the same direction, and remarkably regular. It was inferred that it was unlikely to be caused by relaxation of random process stresses, but was the result of turning memory from the screw. A breaker plate was made and inserted between screw and die; the twisting at once disappeared.

The breaker plate works, of course, by breaking up the plug of polymer containing the aligned 'memory', and reforming it downstream with the alignment fragmented.

Some emphasis has been placed on this matter because it illustrates well the unexpected consequences of the viscoelastic response of these materials to process stresses. We shall meet some related effects in the section on extrusion dies (Section 4.8) where the material response is analysed in more detail.

4.2.4 *Speciality features*

The basic machine described above is used extensively for profile extrusion and for processes which have extrusion as their first stage, e.g. blow moulding and blown film production. However, other uses place extra demands on the extruder and modifications of the basic design have evolved to deal with them.

When the extruder is to be used as a primary mixer special zones having screw flights of changed or even reversed pitch are sometimes employed. The basic single screw extruder is quite a good dispersive mixer but is a poor distributive mixer; these regions of different screw pitch are usually placed beyond the usual metering zone. Their purpose is to induce a sort of quasi-turbulence by mechanical means to improve dispersion. A second metering zone follows, to regularize pressure and temperature before extrusion.

A rather simpler arrangement is to have a *mixing head* beyond the metering zone. At its simplest, this is a 'smear head' (Fig. 4.5). Sometimes its action is improved by means of studs, slots, ribs, etc. to promote mixing.

A more sophisticated recent development for improving the mixing efficiency of an extruder is the *cavity transfer mixer (CTM)* developed at the Rubber and Plastics Research Association, RAPRA. Figure 4.6 shows the

Fig. 4.5 Smear head screw.

Features of a single screw extruder 81

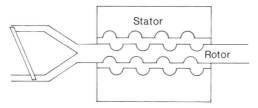

Fig. 4.6 Cavity transfer mixer.

main features of the CTM. This is an example of a static mixing device which depends on cutting and redistributing the melt to achieve mixing. It consists of a cylindrical end, the rotor, attached to the screw, rotating in a cylindrical sleeve, the stator. Rotor and stator both have hemispherical cavities in them, which are staggered with respect to one another. The cavities fill with polymer as the extruder drives it forward and the stream is repeatedly cut and folded as the cavities exchange material.

For some applications it is necessary to have provision for venting volatiles during extrusion. Such machines are provided with a venting port on the barrel. It is, of course, necessary to decompress the melt at this point, to prevent its being expelled from the port. The screw therefore has a decompression region, followed by recompression and a further metering zone (Fig. 4.7).

Sometimes, *vacuum assistance* is advocated for venting. If the volatile matter being expelled is water, this is unnecessary; at a typical extrusion temperature of 250 °C, the water will be steam at ca. 4 MPa, readily expelled

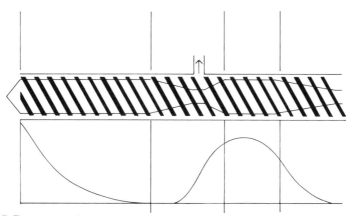

Fig. 4.7 Decompression zone.

82 Extrusion

to atmospheric pressure of 0.1 MPa. Practical opinion varies about the use of venting. Some operators favour it for difficult materials, but others prefer to concentrate on fully adequate predrying of polymer and careful selection of appropriate screw design and processing conditions. It provides an extra degree of processing freedom where a plant has to be versatile, running a variety of different polymers, when of necessity a general purpose screw must be used.

Another modification is the improvement of transport of the polymer granules at the feed end by means of grooves or fins in a prefeed section. This is particularly needed when a mixing head is featured at the other end, because these develop little pressure and the assisted feed counteracts this by developing positive pressure.

Finally, specialized screw designs have been evolved by different manufacturers, often patented, to offer improved performance. Commonly the improvement is in output. Some aspects of these designs will be discussed in Section 4.5 on screw design.

4.3 Flow mechanisms

We have now seen the main physical features of a single screw extruder. How does the machine function to melt and convey the polymer down its length?

4.3.1 *Melting*

As the polymer is conveyed along the screw (see below) a thin film melts at the barrel wall. This is usually by means of conducted heat from the barrel

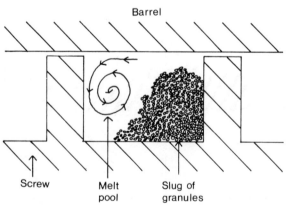

Fig. 4.8 The melting process.

Flow mechanisms   83

heaters, but could be frictional. The screw scrapes off the melted film as it rotates. The molten polymer moves down the front face of the flight to the core and then sweeps up again to establish a rotary motion in front of the leading edge of the flight. Other solid granules or parts of the compacted slug of polymer are swept into the forming 'melt pool'. The process is progressive until all the polymer is melted.

4.3.2 Conveying

To understand the conveying mechanism, consider two extremes.

1. The material sticks to the screw only and slips on the barrel. Under these conditions the screw and material would simply rotate as a solid cylinder and there would be no transport.
2. The material resists rotation in the barrel and slips on the screw. It will now tend to be transported axially, like a normal, deep-channelled, solids-conveying Archimedian screw.

In practice there is friction with both screw and barrel, and this leads to the principal transport mechanism, *drag flow*. This is literally the dragging along by the screw of the melt as the result of the frictional forces, and is the equivalent to the viscous drag between stationary and moving plates separated by a viscous medium. It constitutes the output component for the extruder.

It is opposed by the *pressure flow* component, which is caused by the pressure gradient along the extruder. As we have seen, there is high pressure at the output end, low at the feed end and this pressure gradient opposes the drag flow. It is important to understand that there is no actual flow resulting from the pressure, only an opposition.

The final component in the flow pattern is *leak flow*. There is a finite space between screw and barrel through which material can leak backwards. This is also a pressure-driven flow and of course it also opposes drag flow.

Thus the total flow is the sum of these components

   Total flow = drag flow − pressure flow − leak flow.

4.3.3 Heating and cooling

In high speed machines virtually all the heating is from shearing of the viscous melt. More usually, some heat derives from this source and some from the barrel heaters (see section 2.10); a typical proportioning might be 67/33, friction/conduction. There are also coolers, usually fans to remove excess heat. The whole system is controlled thermostatically to give precise

84   Extrusion

control of melt temperature. The length of the machine is divided into three or four sections to allow variation of temperature for optimum processing.

The practical running condition may be regarded as lying between the extremes of *adiabatic* running, when there would be only heat from viscous dissipation, and *isothermal* running, when the temperature at all points would be the same, with heat being supplied by heaters or removed by coolers to compensate for changes in melt temperature. Real extruders are neither; even so-called adiabatic machines must experience heat losses, and a machine fed with cold stock cannot run isothermally. However, the metering zone may approach isothermal conditions.

4.4 Analysis of flow

In this section we shall briefly derive the standard expression for the output of a single screw extruder. Since this book is not a treatise on extrusion alone the derivation is given in summary. Readers seeking a fuller derivation will find it in one of the standard works listed at the end of the chapter [2], [3].

We shall analyse drag flow, pressure flow, and leak flow and add them to obtain an overall expression for the extruder output.

4.4.1 Drag flow

First, we shall consider flow between a pair of parallel plates and then see how this can be applied to an extruder channel. Figure 4.9 shows parallel plates distance H apart, with a viscous fluid in the space. The moving plate has velocity V_d

- Consider a small finite element of fluid ABCD, distance y from the stationary plate.
- The volume flow rate for this element, dQ, is given by

$$dQ = TV \, dQ. \tag{1}$$

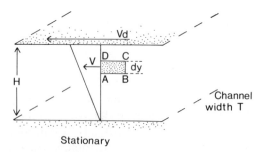

Fig. 4.9 Drag flow between parallel plates.

Analysis of flow 85

- Assume linear velocity gradient and velocity of ABCD is given by

$$V = \frac{V_d y}{H}. \tag{2}$$

Substitute (2) in (1)

$$dQ = \frac{T V_d y \, dy}{H}. \tag{3}$$

- Integrate (3) over channel depth H to find total drag flow, Q_d

$$Q_d = \int \frac{T V_d y \, dy}{H}.$$

$$Q_d = \tfrac{1}{2} T H V_d. \tag{4}$$

Now apply this parallel plate situation to an extruder screw. We can imagine the channel of the screw as a similar channel, except that it has been wound spirally. The appropriate dimensions are shown in Fig. 4.10. The equivalent of the stationary plate is the barrel and of the moving plate is the rotating screw. The finite element is in an equivalent position in the channel. H is now the channel depth. T is the perpendicular distance between flights. N is the screw speed, in revolutions per minute.

The angular movement of the finite element and dimensions can be expressed as functions of the screw flight angle, ϕ

$$V_d = \pi D N \cos \phi$$
$$T = (\pi D \tan \phi - e) \cos \phi$$

Reverting to parallel plates (Equation (4))

$$Q_d = \tfrac{1}{2} T H V_d \tag{4}$$

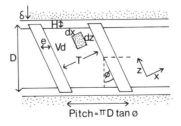

Fig. 4.10 Drag flow in an extruder screw.

and substituting for T and V_d, we have

$$Q_d = \tfrac{1}{2}(\pi D \tan \phi - e) \cos \phi . H . \pi DN \cos \phi$$
$$= \tfrac{1}{2}(\pi D \tan \phi - e)(\pi DN \cos {}^2\phi)H.$$

e is small compared with pitch, so

$$Q_d = \tfrac{1}{2}(\pi D \tan \phi)(\pi DN \cos^2 \phi)H$$
$$= \tfrac{1}{2}\pi^2 D^2 N \tan \phi \cos^2 \phi H$$
$$= \frac{\tfrac{1}{2}\pi^2 D^2 N \sin \phi \cos^2 \phi H}{\cos \phi}$$

$$Q_d = \tfrac{1}{2}\pi^2 D^2 N \sin \phi \cos \phi \; H. \tag{5}$$

Thus we see that drag flow depends on

- screw diameter D^2
- screw speed N
- channel depth H
- helix angle ϕ

i.e. essentially, channel volume × speed, with a factor for helix angle. The helix angle is almost universally fixed at the 'square' angle of 17.66°, i.e. one turn per diameter's length of screw. We shall return to this in the section on screw design (Section 4.5.2).

4.4.2 Pressure flow

The next consideration is to find an expression for the pressure flow. Again, we shall observe a finite fluid element in a channel between parallel plates and then apply the result to an extruder channel. Figure 4.11 shows the viscous fluid element in a pressure gradient. The forces acting on the fluid

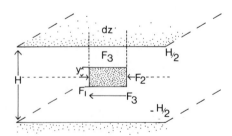

Fig. 4.11 Forces on a fluid element in a pressure gradient.

element are, for unit width

$$F_1 = \left(P + \frac{\partial P}{\partial z}dz\right)2y \quad (\partial P/\partial z\, dz \text{ is the increment caused by pressure gradient along element length } dz)$$

$$F_2 = P2y$$

$$F_3 = \tau\, dz \quad \text{(the viscous reaction)}$$

During steady flow, these are in equilibrium

$$F_1 = F_2 + 2F_3 \tag{6}$$

which reduces to

$$\tau = y\frac{\partial P}{\partial z}. \tag{7}$$

If we assume a Newtonian fluid

$$\tau = \eta\dot{\gamma}$$

$$= \eta\frac{dV}{dy}.$$

Substitute for τ in Equation (7) and assume pressure varies in z direction only

$$\eta\left(\frac{dV}{dy}\right) = y\frac{dP}{dz}$$

$$\frac{dV}{dy} = \frac{1}{\eta}\left(\frac{dP}{dz}\right)y.$$

We can now integrate this to find the velocity V

$$\int_0^V dV = \frac{1}{\eta}\left(\frac{dP}{dz}\right)\int_{H/2}^y y\, dy$$

$$V = \frac{1}{\eta}\left(\frac{dP}{dz}\right)\left(\frac{y^2}{2} - \frac{H^2}{8}\right). \tag{8}$$

From here the volume flow rate dQ is given by

$$dQ = VT\, dy. \tag{9}$$

88   Extrusion

Substitute for V from Equation (8) in Equation (9) and integrate to give the pressure flow Q_P

$$dQ = \frac{1}{\eta}\left(\frac{dP}{dz}\right)\left(\frac{y^2}{2} - \frac{H^2}{8}\right)T\,dy.$$

For $2y$
$$Q_P = 2\int_0^{H/2} \frac{1}{\eta}\left(\frac{dP}{dz}\right) T \left(\frac{y^2}{2} - \frac{H^2}{8}\right) dy$$

$$Q_P = \left(\frac{1}{12\eta}\right)\left(\frac{dP}{dz}\right) TH^3. \tag{10}$$

Now we can apply this expression to an extruder screw channel

$$T = \pi D \tan\phi \cos\phi \quad \text{(neglecting } e\text{)} \tag{11}$$

Now

$$\sin\phi = \frac{dl}{dz}.$$

Therefore

$$\frac{dP}{dz} = \left(\frac{dP}{dl}\right)\sin\phi. \tag{12}$$

Substitute Equations (11) and (12) in Equation (10)

$$Q_P = \frac{\pi D H^3 \sin^2\phi}{12\eta}\left(\frac{dP}{dl}\right). \tag{13}$$

Thus we see pressure flow depends on

- screw dimensions, with dependence on channel depth cubed
- pressure gradient
- fluid viscosity.

4.4.3 *Leak flow*

As we have seen, leak flow is another pressure driven flow component. The geometry is that of a wide slit;

$H = \delta$, the depth of the slit and
$T = \pi D/\cos\phi$, the width of the slit (Fig. 4.12).

However leak flow is small compared with drag flow and pressure flow and may be neglected in finding total flow. It only has practical significance

Analysis of flow

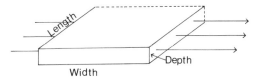

Fig. 4.12 Leak flow through a wide slit.

in badly worn machines where the clearance between screw and barrel becomes large.

4.4.4 Total flow

We can now easily find the total output flow Q by summing the expressions for drag flow and pressure flow

$$Q = Q_d + Q_p$$

$$Q = \tfrac{1}{2}\pi^2 D^2 NH \sin \phi \cos \phi - \left(\frac{\pi DH^3 \sin^2 \phi P}{12\eta l} \right).$$

This is a somewhat cumbersome expression, which for practical purposes is simplified. For a given extruder, l, D, H, ϕ are all fixed, and we can thus say

$$Q = \alpha N - \left(\frac{\beta P}{\eta} \right).$$

The practical variables for operating the extruder are

- screw speed N
- head pressure P
- melt viscosity.

4.4.5 Influence of polymer properties

The expression for extruder output above is obviously something of a simplification. The two important factors missing are (a) the non-Newtonian rheology of most polymer melts and (b) their frictional properties. As we have seen, the material will not progress unless there are frictional forces. The importance of frictional drag at the barrel wall may be compared with the tightening of a nut on a bolt. Unless the nut is held fast, the bolt will not tighten. In the same way the frictional drag prevents the melt from simply rotating with the screw. However, unlike the firmly held nut, the frictional force is variable, being dependent on polymer type and conditions. The more friction there is, the less the tendency to rotate with the screw, hence

90 Extrusion

the development of longer screws over the years, the increased barrel surface area offering more overall frictional force.

Next, consider a little more critically the standard expression for drag flow developed above. It was based on a number of simplifying assumptions

- The melt behaved as a Newtonian fluid;
- Thus, the viscosity is the same at all points;
- There is a velocity gradient from zero at the screw to maximum at the barrel wall.

These assumptions lead to the factor ½ in the drag flow equation

$$Q_d = \tfrac{1}{2}(\pi^2 D^2 NH \sin \phi \cos \phi)$$

i.e. it is an integration constant, as we saw above.

But we know that these materials are non-Newtonian fluids; also the velocity cannot be zero at the screw because this would imply a stagnant layer which would degrade. The material must slide on the surfaces and the sliding characteristics are described by the relevant coefficients of friction. These vary over a wide range for different polymers; furthermore, they are highly dependent on temperature. The breadth of this variation is indicated by a few illustrative values in Table 4.1. The drag flow condition is thus better represented if the factor ½ is replaced by a more general one, F, which incorporates frictional and viscous response of the polymer

$$Q_d = F(\pi^2 D^2 NH \sin \phi \cos \phi).$$

Under idealized conditions, $F = \tfrac{1}{2}$, but with low barrel friction, high screw

Table 4.1 Coefficients of friction, various countersurfaces

Polymer	Coefficient of friction (λ)
PTFE	0.04–0.15
LDPE	0.30–0.80
HDPE	0.08–0.20
PP	0.67
PS	0.33–0.50
PMMA	0.25–0.50
Nylon	0.15–0.40
PVC	0.20–0.90
SBR	0.50–3.0
NR	0.50–3.0

After Hall [4].

friction and low apparent viscosity, $F \to 0$ and the drag flow diminishes towards zero. Under these conditions successful operation of the extruder becomes very difficult; the drag flow equation indicates that a large diameter screw (high D) running at high speed (high N) is needed to maximize output, but this tends to increase shear rate

$$\dot{\gamma} = \frac{\pi DN}{H}$$

which lowers viscosity even further and hence decreases F. A further complication is the dependence of the frictional coefficient on temperature. This is different for each material and is a major factor in determining the ease with which it will run in an extruder. For example for LDPE, λ at $100° \simeq 1$, and at $250° \simeq 0.055$ [5]. Thus it slides well against the hot barrel wall. Friction decreases with temperature providing a self-regulation of frictional heat development. This material is easily processed in a single screw extruder.

Contrast this with unplasticized PVC, which shows the opposite behaviour. For PVC, λ at $100° \simeq 0.06$, and at $150° \simeq 1.00$. Now the friction necessary to maintain output is easily exceeded; as the temperature rises to obtain a satisfactory melt, the frictional response increases and overheating results from frictional heating, and consequent degradation readily occurs. PVC is a polymer notorious for thermal instability; it requires quite elaborate heat stabilizing systems as additives under good processing conditions. Thus, it is a difficult polymer to run in a single screw extruder,

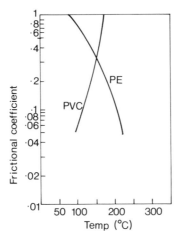

Fig. 4.13 Comparison of frictional properties of polyethylene and PVC (after Jacobi).

92 Extrusion

and is more often extruded in twin screw machines. These two polymers are compared in Fig. 4.13.

4.5 Some aspects of screw design

Two important aspects of extruder performance concern the design of screws for these machines. These are the efficiency of melting and the output rating of the extruder.

4.5.1 *Melting efficiency*

Consider again the melting process as observed in a single screw extruder. This was described in Section 4.3.1 and Fig. 4.8. This part of the process, in which the solid granules are progressively melted into the accumulating melt pool, has deficiencies which have led to many developments in design. Some of these trends have been described by Lovegrove [6] and are outlined below.

The melting process is efficient early, but as melting proceeds the proportion of channel occupied by solids diminishes so that contact between solids and the hot barrel is reduced. The 'tongue' of solids breaks up and solid particles become reliant on heat from the surrounding melt to complete their melting. This is a source of inhomogeneity and, of course, it is one of the functions of the metering zone to even these out. The effects of variations in screw characteristics in this context may be summarized.

1. Deeper channel: conveys more material but takes longer to complete melting;
2. Fast running: maximizes output but solids persist further along the screw;
3. Shallower channel: can help fast running to increase output because of more efficient melting but the danger is that the resultant high shear might lead to overheating.

One of the developments to improve melting efficiency is the *barrier flight* screw. An extra flight is provided, separate from that containing the solid granules. Its clearance within the barrel is smaller than granule size. As melting begins the molten polymer can move into the new channel and solid and melt are thus separated. The normal and barrier flights have different helix angles; the solids channel thus starts wide and narrows, the melt channel vice versa. The early efficiency of melting is thus continued as the developing melt is strained off and the channel volume decreases to accommodate the diminishing volume of solids. The inventor of the barrier flight screw was Maillefer and these screws are often referred to generically by this name, although other makes and variants exist. Figure 4.14 shows the principle of the Maillefer screw.

Some aspects of screw design

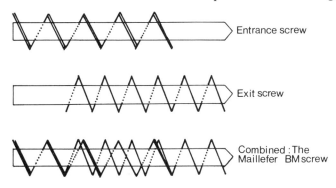

Fig. 4.14 Principle of the Maillefer screw.

Another important approach to improved melting is the use of mixing devices prior to the metering zone; mention of these has already been made in Section 4.2.4, in connection with the use of extruders as mixers. The mixing action also improves melting rate simply by 'stirring up' the material. When extended back to the region where 50% or more of the material is still solid, the melting efficiency is markedly improved. However, such mixing elements fail to develop the required output pressure and consequently grooved feed sections are built in to develop the pressure. A comparison of the performance of various designs is seen in Table 4.2. This comparison shows how output improves as the result of more efficient melting. We can see also that the high pressure feed design consumes more power than the rival barrier flight; this is because of the need to cool its HP grooved feed section – 14–20% of the power is used for this; the pay off, however, is in the output. This is an area of very active development by extruder manufacturers, with new models continually appearing.

Table 4.2 Comparison of different screw designs 90 mm dia. extruder, HDPE, max., temp. 220 °C

Screw type	Rating ($kg\,h^{-1}\,r.p.m.^{-1}$)	Maximum r.p.m.	Output ($kg\,h^{-1}$)	Power ($kWh\,kg^{-1}$)
Conventional	2	75	150	–
Simple mixing screw	2.2	82	180	–
Barrier flight	2.67	100	267	0.2
High press. feed design	3.5	120	>400	0.23–0.27

After Lovegrove [6].

4.5.2 Optimum helix angle

One of the mysteries surrounding extruder screw design is the apparent sanctity of the 'square' helix angle. Virtually all screws are made in this way, one turn of the helix occurring in one diameter's length. The competing requirements are a steep angle to resist back pressure flow and a shallow angle to provide the least tortuous path for drag flow (Fig. 4.15). What is the

Fig. 4.15 Steep vs. shallow helix angle: the steep angle resists back pressure flow; the shallow angle provides least tortuous drag flow path.

explanation for this almost universally accepted value of 17.66°, and can it be justified in the light of modern concepts of polymer behaviour? We can arrive at a rationale for the square angle in a straightforward manner from consideration of volumetric efficiency, using the expression for drag flow already discussed.

The 'ideal' axial velocity, V_a can be described as

$$V_a = \text{screw pitch} \times \text{screw speed}$$
$$= \pi D \tan \phi \times N.$$

The velocity component parallel to the screw flight is V_d

$$V_d = V_a/\sin \phi$$
$$= (\pi DN \tan \phi)/\sin \phi \quad \text{(Fig. 4.15)}.$$

The 'ideal' output, Q_i is given by

$$Q_i = V_d \times \text{cross-section of channel}$$
$$= \frac{\pi DN \tan \phi}{\sin \phi} (\pi DH \tan \phi) \cos \phi$$
$$Q_i = \pi^2 D^2 HN \tan \phi.$$

'Efficiency' $= \dfrac{Q_{max}}{Q_i} = \dfrac{\frac{1}{2}\pi^2 D^2 NH \sin\phi \cos\phi}{\pi^2 D^2 NH \tan\phi} = \frac{1}{2}\cos^2 \phi.$

Thus, the volumetric efficiency, at least of drag flow, depends only on ϕ. The square angle of 17.66° gives an efficiency of 45.4%. If ϕ is steepened to

10° the efficiency rises to only 48.5%, and there is diminished ability to develop useful head pressure at the die. At higher values of ϕ the efficiency falls off sharply. The general compromise has been to adopt the intermediate square angle. However, Rauwendaal [7] has shown that the optimum helix angle for melt conveying depends, for a power law fluid, on the flow behaviour index (see section 2.6).

$$\sin \phi_{opt} = \left(\frac{n}{2n+2}\right)^{1/2} + \hat{w}\left(\frac{n+2}{4n}\right)$$

where $\hat{w}$ is the reduced flight width in

$$\hat{w} = \frac{pw}{\pi D}$$

where p is the number of flights, w is the perpendicular flight width, and D is the screw outer diameter.

If this expression is used to determine the helix angle in one or two examples, we find values rather different from the square angle value of 17.66°. In Table 4.3 values of n for a number of polymers are used to find ϕ_{opt}. Three values of w are used, 0 (i.e. the flight width is neglected), 0.1 and 0.2.

Table 4.3 Optimum helix angle and flow behaviour index

Polymer	n	ϕ_{opt} (°) $\hat{w} = 0$	$\hat{w} = 0.1$	$\hat{w} = 0.2$
PMMA, ABS	0.25	18	33	50
PS, PVC	0.3	20	32	46
Polypropylene	0.35	21	32	44
HDPE	0.5	24	32	41
PET	0.6	26	33	41
Polycarbonate	0.7	27	33	40
Nylon 6,6	0.75	28	34	40
Newtonian fluid	1.0	30	35	41

This illustrates the empirical way in which process design evolved. If the flight width is neglected, many highly non-Newtonian polymers approach the square angle value of ca. 18°, and experience with early polymers would reflect this empirically; this is then supported by the simple approach to efficiency which assumes Newtonian viscosity. Other aspects of extruder and screw design have been quite drastically altered over the years, as we have noted above, but, strangely, helix angle has seldom been changed,

96 Extrusion

even experimentally. However, this situation is beginning to yield and a new type of screw has emerged for improved output of linear low density polyethylene, LLDPE, often called the 'LL screw'. It has a helix angle in the metering section of 27.5° and a narrow flight width. These factors lead to a power saving of 35–40% for a given output [8]. The data in Table 4.3 suggests that the LL helix angle should be more productive with other polymers as well. There is a considerable inertia before such design changes become commonplace, largely because of the high capital cost of such equipment, but it seems likely that in the future a more imaginative approach to screw geometry will be seen, especially with regard to helix angle.

4.5.3 Other design considerations [9]

A few other general considerations are worth mentioning. Is the screw likely to sag under gravity? It is a cantilever held only at its drive end. Normal screw/barrel clearances ($\simeq 0.2$ mm) are exceeded as the result of sag when the L/D ratio is more than ca. 10. Physical support from the polymer is thus important to prevent contact, and prolonged empty running is obviously unwise.

Buckling caused by high head pressure is a theoretical possibility with long screws ($L/D > 20$), but the lateral forces are in practice easily supported by the melt.

Whirling is the phenomenon which causes rotating shafts such as stirrers and drills to become 'whippy' and unstable at rotational speeds which allow resonant vibrations to develop. They are unimportant in extruder screw behaviour because the speed would have to exceed ca. 2000 r.p.m. for their appearance; this is well outside normal practice.

A more important source of screw–barrel contact and hence wear is deflection of the screw caused by pressure difference side to side, i.e. across the diameter. Such effects are most likely at points of highest pressure, e.g. at the end of the compression zone. If the compression ratio is too high the solid bed can plug the channel, which increases the pressure locally and deflects the screw into the barrel wall. The result will be severe wear, expensive and difficult to repair. A pressure difference ΔP across the diameter D will cause a lateral force F over length L

$$F = \Delta P \times D \times L.$$

For $D = L = 150$ mm, $\Delta P = 1$ MPa, F is 22.5 kN ($\simeq 5000$ lbf). As Rauwendaal points out [9], such forces readily cause substantial wear and furthermore are difficult to diagnose because the wear often develops gradually. Even so, the uneven melting resulting from too high a compression ratio and the consequent pressure surging, which should be apparent in irregularities in the emerging extrudate, should be regarded as danger signals. If solid particles really penetrate badly they may even appear trapped at the screen pack.

4.6 Twin screw extruders

As we have seen, the single screw extruder, although the workhorse of the industry, possesses deficiencies in certain applications. Among the most prominent of these is the difficulty of extruding heat sensitive polymers like PVC, especially where the frictional properties of the polymer aggravate the problem. For such sectors of the extrusion industry twin-screw extruders offer a better performance. One of these sectors, which has shown spectacular growth in recent years, is the extrusion of window frames and associated products, e.g. frames for patio doors, main house doors, etc., in uPVC, i.e. unplasticized or rigid PVC. These profiles are usually extruded directly from powder blends, which are notoriously difficult to run in conventional extrusion and the success of this industry has depended in no small measure on the use of twin-screw machines; pelletized, pre-compounded PVC, on the other hand, is readily run on single screw machines.

What do the twin-screw machines offer, and how does their action differ from that of the single-screw extruder? Before we can answer this question, it is necessary to recognize the different categories of twin-screw extruders.

4.6.1 *Categories of twin-screw extruders*

Twin-screw extruders divide into *co-rotating* and *counter-rotating* types. As the names indicate, the difference is in whether the two screws rotate in the same or in opposite directions, i.e. both clockwise or counterclockwise, or one in each sense.

The next division is determined by whether the two screws mesh with each other or not; they are described as *meshing* or *non-meshing*. The non-meshing types consist of essentially two single screws side by side and work in a similar manner to single-screw machines; they are not true twin-screws and are better described as 'double screws'.

Within the meshing (or intermeshing) types, there is a further subdivision into *conjugated* and *non-conjugated* machines, depending on whether the meshing flights fully occupy the channels in the meshing area.

4.6.2 *Melt conveying in twin-screw extruders*

The action of twin-screw extruders is less well worked out than is the case for their single-screw counterparts. A particularly useful account, however, has

98 Extrusion

been produced by Martelli [10]. A summary of the action of these machines, based on this work is given below.

Twin-screw extruders act as positive displacement pumps with little dependence on friction, and this is the main reason for their choice for heat sensitive materials. Counter-rotating machines, if conjugated may have no passage at all for material to move around the screws; it must move axially towards the die end. Likewise, co-rotating machines will have no passage around each screw and only a small and tortuous one round both, also leading to positive axial flow. Thus we see that L/D ratio is unimportant for propulsion and so there is no long metering zone. The length must, of course, be sufficient for proper melting of the polymer. Also because of the positive pumping action the rate of feed is not critical in maintaining output pressure. The conveying is not by drag flow, and this allows good control of shear rate and hence temperature, especially important with polymers like PVC.

Let us look in a little more detail at the action of these machines.

4.6.3 *Action of counter-rotating screws*

When the screws rotate points which are next to one another on the two screws remain so as they move through the intermeshing region. The plane on which they move is inclined to the screw by the pitch angle, but each point on each screw moves on a plane at the same inclination so that there is no interference. This means that any shape of flight may be used.

When flights of rectangular section are made as thick as possible they fill the opposing channel where they intermesh. Thus they leave no passage between the screws, i.e. they are fully conjugated on the plane of the axes of the screws. The polymer is thus enclosed in C-shaped chambers round each screw, which leads to positive pumping but restricts mixing. The pumping action moves the material towards the output end like a nut on a bolt. Although drag action is unimportant for output it is of course still occurring; in the C-shaped chamber the material is dragged to the intermeshing point, where a mill-like action occurs. This causes a great increase in pressure as the material squeezes through the mechanical clearances under high shear. An adverse effect of this pressure is that it tends to sunder the screws, causing screw and barrel wear.

The flights can be made thinner, so that they only partly impede the individual screw behaviour, and although this reduces the pressure problem it also rather defeats the twin-screw concept because there is then no conjugation. Thus, rectangular flight sections are not very satisfactory. Flights of trapezoidal shape (Fig. 4.16), are rather more satisfactory; they allow conjugation in the plane of the axes and at the same time some space for material to move between channels. Flights and channels of this form permit some mixing action, screw to screw, but at the plane of conjugation

Twin screw extruders 99

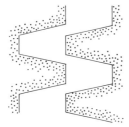

Fig. 4.16 Trapezoidal flight and channel forms.

the problem of high shear and pressure remains. Extruders of this general type are often called 'CICT', (closely intermeshing counter-rotating). In practice, compromises have to be made between perfect conjugation with its small clearances and self cleaning, and some reduction in conjugation through larger clearances to reduce shear and pressure. These factors show up later in assessing the output efficiency of these machines.

4.6.4 *Action of co-rotating screws*

The action of co-rotating screws is quite different from that in counter-rotating machines. Now, two points adjacent to start with become separated as the screws rotate, because one flight goes forward and the other rearward. Rectangular flights are impossible; the simplest are triangular or trapezoidal (Fig. 4.17). This configuration has gaps which allow circulation of polymer between the screws in a figure-of-eight pattern. Now, when the melt arrives at the conjuction it experiences no milling action. Both screws are moving in the same sense at this point, with the result that material can be transferred to the opposite screw in the figure-of-eight path. This tends to reduce the positive pumping efficiency. However, the fraction of material involved in this interchange is quite small, with two important consequences; (a) the positive pumping action is quite good, though less effective than in the CICT

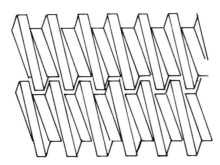

Fig. 4.17 Trapezoidal intermeshing co-rotating screws.

case, and (b) there is a problem with sundering pressure between the screws. These machines have the designation 'CICO' (closely intermeshing co-rotating). A more commonly used design is the so-called 'self-wiping' configuration, 'CSCO' (closely self-wiping co-rotating). The flight profile here is sinusoidal, like the thread of a bolt (Fig. 4.18). This configuration is

Fig. 4.18 Sinusoidal self-wiping screws.

characterized by quite large figure-of-eight passages for screw-to-screw interchange. Mixing is now quite effective and the pressure between screws is practically eliminated, which allows high rotational speeds. However, the loss is in pumping efficiency and these machines are not well suited to profile extrusion, being more applied as mixers.

4.6.5 *Output of twin-screw extruders*

As we have seen, the mechanical action of twin-screw extruders is primarily a positive pumping one. Drag flow, with its dependence on viscosity and frictional properties, is only of minor importance. The equations which express their output therefore derive from their geometry, and we find no terms for viscosity, friction or pressure. The derivation of expressions for output is given by Martelli [11]. In essence, it depends on finding, for counter-rotating screws, the volume of the C-shaped chambers, and for co-rotating screws that of the figure-of-eight passages between screws. What emerges is

For counter-rotating screws, $Q_c = \pi DH \sin \phi \, (\pi D - \sqrt{2DH})N$.
For co-rotating screws, $Q_c = \frac{1}{2}\pi^2 D^2 NH \tan \phi$.

Both expressions maximize at $\phi = 90°$. They depend on screw dimensions and N, the screw speed only. Thus we can simplify them for comparison with the simplified expression for single-screw machines (Section 4.4.4) as follows

> For a single-screw machine $Q = \alpha N - \beta P$
> For a twin-screw machine $Q_c = \psi N$ (counter-rotating)
> $Q_c = \omega N$ (co-rotating)

where α, β, ψ and ω are machine constants.

At its simplest, the output of a given twin-screw machine depends only on its speed. However, these simple expressions require modification to account for inefficiencies. As we saw in the description of the action of conjugated twin-screws, some inefficiencies have to be allowed in the compromises between maximum pumping and the problems of sideways pressure on screws. There are also, of course, the inefficiencies deriving from normal machine clearances. The expressions above for Q_c are for theoretical maxima. To account for the leakage inefficiencies a quantity q may be subtracted

$$Q_{max} = Q_c - q$$

q is expressed as a fraction of Q_c. Martelli quotes values of q

> q for counter-rotating screws $= 0.65 - 0.5\, Q_c$
> q for co-rotating screws $= 0.15 - 0.10\, Q_c$.

This implies greater pumping efficiency with co-rotating screws. However, as we have seen, extruders with self-wiping co-rotating screws have only poor pumping efficiency and Martelli's values would seem to apply to the type with trapezoidal flights – the CICO pattern.

A further consideration is that it is the normal practice with twin-screw extruders to run with the screws only partially filled. Q_{max} refers to the condition with full screws, fed on demand. Controlled feeding allows operation with partial filling, which is preferred because it relies solely on the pumping action of the extruder, with minimal dependence on shearing the melt. This is, as we have seen, the main attraction of these machines for heat sensitive polymers. The actual output under these running conditions is Q and the ratio Q/Q_{max} is the *filling ratio* for the machine. As long as Q_{max} exceeds the required output the machine will perform the task. When N (the machine speed) is increased Q_c increases but so does q. Q_{max} thus increases with speed but not linearly.

It will be clear from this account that twin-screw extruders are complex in their action. There are diverging views about the respective merits of co- and counter-rotating designs, and the many design variants available from different manufacturers. Most manufacturers of rigid PVC window profiles and pipes use counter-rotating extruders but some do favour co-rotating designs. In theory, the co-rotating configuration is superior in many ways because it avoids the milling action at conjunction and it may well be that the preference will swing in this direction in the future.

4.7 Extruder and die characteristics

The product of extrusion is realized by forcing the melt through a shaped die to give a profiled, continuous 'extrudate'. The interaction of the extruder and its die can be understood by looking at their respective 'characteristics'. These concepts are developed as follows.

1. If there were to be no pressure build-up, for example, no breaker plate or die, the output would be at its maximum, Q_{max}. We can use the drag flow ideal equation
$$Q = Q_{max} = \tfrac{1}{2} \pi^2 D^2 NH \sin \phi \cos \phi.$$
2. If there is maximum resistance and $Q = 0$, and we can equate the drag flow and pressure flow expressions
$$P = P_{max}$$
i.e. $\tfrac{1}{2}\pi^2 D^2 NH \sin\phi \cos\phi = \dfrac{\pi DH^3 \sin^2 \phi P}{12\eta l}$

$$P_{max} = \frac{6\pi Dl N \eta}{H^2 \tan \phi}.$$

These points represent the extremes in the diagram of the *extruder or screw characteristic* (Fig. 4.19).

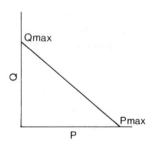

Fig. 4.19 The extruder or screw characteristic.

3. A die at the extruder outlet requires head pressure to function; the pressure is needed simply to force the melt through the die. The *die characteristic* is thus opposite in form; the maximum output will result from maximum pressure. This can be added to the diagram to give Fig. 4.20. The intersection point is the *operating point*, where the optimum operating conditions are.

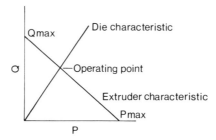

Fig. 4.20 Interaction of screw and die characteristics.

4. The output of individual dies depends, obviously on their shapes. In general

$$Q = KP$$

where K is the shape-dependent factor. For cylindrical capillary dies

$$K = \frac{\pi R^4}{8\eta L}$$

where R = radius and L = length.

5. The positions of the lines in Fig. 4.20 will be changed by changes in operating conditions. If N, the extruder speed, is increased the extruder characteristic moves up. If the die dimensions are changed, e.g. the radius R of the cylinder is increased, the slope of the die characteristic is increased. These effects are developed in a little more detail below.

The output equation for the extruder is, as we have seen

$$Q = \alpha N - \left(\frac{\beta P}{\eta}\right).$$

α and β/η are constants for the total system including the polymer. The equation is thus that for a straight line with negative slope. A family of parallel screw, or extruder characteristics can be drawn for various screw speeds, N (Fig. 4.21).

The slopes of these lines are determined by the β/η term, which is the term $\pi DH^3 \sin^2 \phi / \eta l$. Thus, the slope depends on H^3, the cube of flight depth, $\sin^2 \phi$, the square of helix angle (probably 17.66°), and $1/l$, the reciprocal of screw length. Increases in these quantities flatten the screw characteristic. For a steep characteristic, the system is sensitive to pressure changes – a small pressure increase at the head deceases output sharply. For a flat characteristic, the output is not markedly changed if the head changes.

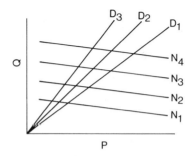

Fig. 4.21 Family of screw characteristics, with various die intersections.

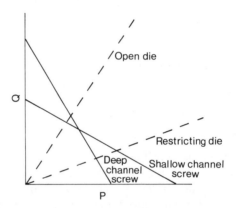

Fig. 4.22 Different matches of screw and die characteristics.

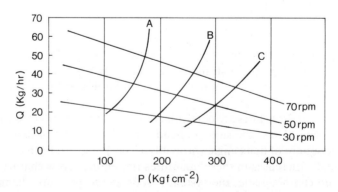

Fig. 4.23 Output characteristics for a 60 mm extruder (60 mm extruder, short compression screw, polyethylene, MFI 0.5, three different die characteristics).

The extrusion die   105

In practice different matches of screw and die are appropriate (Fig. 4.22). The intersection points show that with an open die the best results would be obtained with a deep channelled screw, whereas, with a restricting die, a shallow channel would work better. Figure 4.23 shows diagrammatically some output characteristics (after Fisher [12]). Notice that in the practical example in Fig. 4.23, the die characteristics are curved. This increase in output means increased shear rate and hence lowered apparent viscosity.

4.8 The extrusion die

4.8.1 *Basic flow patterns*

Design of a satisfactory extrusion die is a difficult and critical matter. Although some principles are well established, and the behaviour of melts in constricted channels is being more and more understood, there remains an element of design and construction which relies on experience and art.

In this book our main concern is with the behaviour of polymers during processing, rather than engineering design of plant, and it is this aspect of extrusion dies which is the subject of this section. Figure 4.24 shows some conjectural patterns for a die for extruding rod, which we can use to examine a few of the principles.

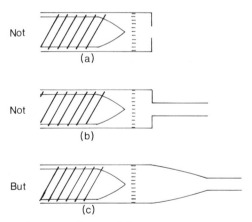

Fig. 4.24 Conjectural patterns for a rod die.

Why will a zero-length profile, like that shown in part (a) of Fig. 4.24, not serve, and why are the tapers of part (c) needed instead of the abrupt changes of (a) and (b)? The answers lie in the need to maintain *laminar flow* in the melt. If the changes are abrupt as in (a) and (b), 'dead spots' occur in

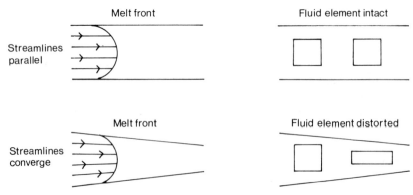

Fig. 4.25 Parallel and converging flows.

the corners where the melt circulates like a backwater and this leads to an extrudate with uneven heat and shear history.

Also, in any converging flow, there are *tensile* as well as *shear* forces. The tensile properties were mentioned in Section 2.4, and are important in several polymer processing routes. Their relevance in die behaviour is simply illustrated in Fig. 4.25. In the parallel flow arrangement, the streamlines are seen to be parallel: a fluid element remains intact as it proceeds along the channel. In the case of converging flow, the streamlines converge and the fluid element becomes distorted as the result of developing tensile stress, i.e. the melt is stretching as it narrows down towards the exit.

The simple rule for deciding whether there will be tensile as well as shear forces operating is to ask whether the streamlines are parallel; if they are there is a simple shear, but if they are not there is a tensile component. How does this affect the die profile design?

4.8.2 *Die entry effects*

If the tensile stresses become large, as they would in parts (a) and (b) of Fig. 4.24, they can actually exceed the tensile strength of the melt, which is usually about 10^6 Nm^{-2}. When this happens the streamlines become not only chaotic but discontinuous; the desirable smooth, laminar flow is lost completely. The extrudate emerging from the die exit will be of irregular shape. Instead of a smooth rod in our example, a jagged strip will emerge (Fig. 4.26). This is the phenomenon known as *melt fracture*. It would occur very readily with a die like that in (a) of Fig. 4.24, which is simply an orifice with zero length. The extended parallel section, known as the *die land*, in Fig. 4.24(b) will be of little value in remedying this situation; the damage has already been done at the die entrance.

In Fig. 4.24(c) is seen the principle of the correct approach; the die

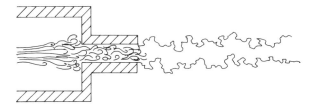

Fig. 4.26 Melt fracture.

entrance is tapered. The effect of this is to

(i) eliminate the dead spots in the corners, hence maintaining a steady heat and shear history
(ii) minimize the development of tensile stresses, and hence minimize distortion of the streamlines.

The long die land is now valuable in steadying towards a parallel configuration of low lines before emergence from the die exit. It also extends the process time which helps to eliminate memory of earlier processing, e.g. the screw turning memory, or a constriction or change in direction in the flow path. The smoother the streamlines, the faster it will be possible to run the process and the better the product will be. If due account is not taken of the effects of 'memory', the result is likely to be distorted extrudate (see Section 4.2.3).

How can we quantify this? The situation is quite complex, because the shear rates, and hence the apparent viscosities, vary for different stages of the process, and these stages are also of different durations. An interesting technique has been derived by M. Reiner [16]. It involves finding the *Deborah Number*, N_{deb}*, as follows.

The characteristic timescale for which a melt has memory is related to its *relaxation time*. Full accounts of viscoelastic behaviour and relaxation times, as observed in polymer melts may be found in more specialized texts [13, 14]. In outline, relaxation time is found from the *viscosity* and the *elastic modulus*; these are the quantities that describe its viscous and elastic responses to an applied stress, and their ratio gives us a *relaxation* or *natural* time for the material

$$\text{Relaxation time} = \frac{\text{viscosity}}{\text{modulus}} = \frac{N s \times m^2}{m^2 \times N} = s.$$

We need to find this natural time for the material under the processing conditions in use. It is then compared with the timescale of the process, and

---

* Judges V: The song of Deborah and Barak; 'The mountains melted from before the Lord.'

the result of this is the Deborah Number, N_{deb}

$$N_{deb} = \frac{\text{Relaxation time of material, in process}}{\text{Timescale of process}}.$$

If $N_{deb} > 1$ the process is dominantly elastic. If $N_{deb} < 1$ the process is predominantly viscous.

F.N. Cogswell has provided a convenient example [15], although our interpretation differs somewhat. Figure 4.27 represents a melt flowing round a bend and then through a relatively narrow die. The timescale from

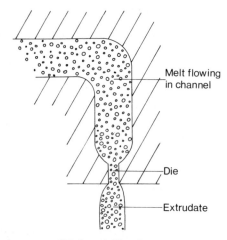

Fig. 4.27 Process memory and Deborah Number.

the bend to the die might be, say 10 s. The flow in this region involves a relatively low stress, giving a relatively high viscosity, say, 10^5 N s m^{-2}. At the same time, the elastic modulus *increases* with increasing stress. Now use the Deborah Number concept to describe what happens.

1. Flow from bend to die

   low stress: $\dfrac{\text{viscosity}}{\text{modulus}} = \dfrac{10^5 \, \text{N s m}^{-2}}{10^3 \, \text{N m}^{-2}}$

   relaxation time $= \dfrac{10^5}{10^3} = 100 \, \text{s}$

   timescale of process $= 10 \, \text{s}$

   $N_{deb} = \dfrac{100}{10} = 10$

   i.e. this process is predominantly *elastic*.

2. Flow through die

high stress: $\dfrac{\text{viscosity}}{\text{modulus}} = \dfrac{10^3 \,\text{N s m}^{-2}}{10^5 \,\text{N m}^{-2}}$

relaxation time $= \dfrac{10^3}{10^5} = 0.01\,\text{s}$

timescale of process $= 0.1\,\text{s}$

$N_{\text{deb}} = \dfrac{0.01}{0.1} = 0.1$

i.e. this process is predominantly *viscous*.

Now we need to interpret this result in terms of the physical behaviour of the extrudate. Instinct probably leads one to expect that the high shear, viscous process 'cancels out' all that has gone before; everything has been put back into the melting pot, almost literally. However instinct is wrong. The extrudate will curl! (see section 4.2.3).

It does so because the process in the bend, though of a long time-scale, is elastic, i.e. the chains are not permanently realigned. The subsequent short-timescale deformation in the die, although developing high shear, does not remove this elastic distortion, which can be released as the extrudate leaves the restriction of the die with the result that it curls.

In practice, it is often difficult to know what the shear rates will be with any precision, or to put values on the viscosity and modulus; the contribution of the tensile component may not be known. In consequence, the tool designer and maker have to resort to experience and an instinctive feeling for what will work. Much trial and error goes into the establishment of satisfactory dies, especially for the extrusion of complex profiles, and a calculation for design purposes involving Deborah Number would be unlikely. The value of the above use of Deborah Number is that it helps towards an understanding of the sometimes unexpected behaviour of polymer melts.

Melt fracture and process memory are examples of die entry phenomena, and the die itself has only minimal effect in remedying them. What of the die exit?

4.8.3 Die exit instabilities

The most common defect found at the die exit is known as *sharkskin*. It is a roughening of the surface of the extrudate. This is another effect caused by tensile stresses, in the following way.

The melt, as it proceeds along the die channel, has a velocity profile, with maximum velocity at the centre and zero velocity at the wall. As it leaves the die lips, the material at the wall has to accelerate to the velocity at which the extrudate is leaving the die. This generates tensile stress, and, if the stress exceeds the tensile strength, the surface ruptures causing the visual defect. As expected, low modulus, high elongation materials are least affected; structured, highly filled, low elasticity materials most easily show sharkskin.

If the conditions causing sharkskin become more intense, e.g. pressure at the extruder becomes excessive, or the die temperature drops, the effect deteriorates towards a coarser-grained appearance, often called *orange peel*. Eventually, the recovery of the tensile forces becomes wholesale and the whole extrudate 'snaps back'. The result is *bambooing*, so called because the extrudate resembles the appearance of bamboo (Fig. 4.28). Extra heating of the die will often help to remedy these defects, by thermally relaxing the stresses and lowering viscosity.

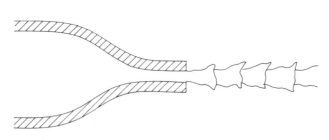

Fig. 4.28 'Bambooing' at a die.

It is a common misconception that sharkskin is a sort of mild melt fracture, and it is true that the onset of sharkskin, as the extruder speed is increased, may be followed by the complete break-up of the extrudate that characterizes melt fracture if the speed is raised further. However, as we have seen, the two phenomena have different origins, so that the action taken to reduce one may not prevent the other.

4.8.4 *Die swell*

Die swell is the effect in which the polymer swells as it leaves the die. The result is an extrudate which differs in its dimensions from those of the die orifice. Thus an extruded rod would be of larger diameter and a pipe would have thicker walls, i.e. increased o.d. decreased i.d. (Fig. 4.29). Die swell is another result of the elastic component in the overall response of a polymer melt to stress. It results from recovery of the elastic deformation as the extrudate leaves the constraint of the die channel and before it freezes.

References and Further reading

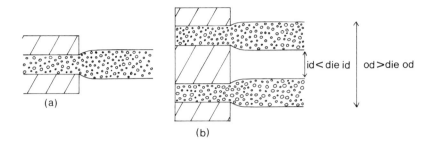

Fig. 4.29 Die swell in (a) rod and (b) pipe.

References

1. Rauwendaal, C. (1986) *Polymer Extrusion*. Hanser, New York, p. 390.
2. Crawford, R.J. (1981) *Plastics Engineering*. Pergamon, Oxford.
3. Stevens, M.J. (1985) *Extruder Principles and Operation*. Elsevier Applied Science, Barking, U.K.
4. Hall, C. (1981) *Polymer Materials*. Macmillan, London.
5. Jacobi, H.R. (1963) *Screw Extrusion of Plastics*. Iliffe, London.
6. Lovegrove, J.G.A. (1982) *Plastics & Rubber Weekly,* 5 June.
7. Rauwendaal, C. (1986) *Polymer Extrusion*. Hanser, New York, p. 361.
8. Rauwendaal, C. (1986) *Polymer Extrusion*. Hanser, New York, p. 387.
9. Rauwendaal, C. (1986) *Polymer Extrusion*. Hanser, New York, p. 355.
10. Martelli, F.G. (1983) *Twin Screw Extruders, A Basic Understanding*. Van Nostrand, Reinhold, New York.
11. Martelli, F.G. (1983) *Twin Screw Extruders, A Basic Understanding*. Van Nostrand, Reinhold, New York, pp. 23–40.
12. Fisher, E. (1964) *Extrusion of Plastics*. Iliffe, London.
13. Nielsen, L.E. (1962) *Mechanical Properties of Solid Polymers*. Van Nostrand, Reinhold, New York.
14. Hall, C. (1981) *Polymer Materials*. Macmillan, London.
15. Cogswell, F.N. (1981) *Polymer Melt Rheology*. Godwin, London.
16. Reiner, M. (1960) *Deformation, Fractures and Flow* (ed. H.K. Lewis).

Further reading

Rauwendaal, C. (1986) *Polymer Extrusion*. Hanser, New York.

5
Extrusion-based processes

In Chapter 4 the action of screw extruders and their dies was studied. This chapter describes some of the most important production processes in which the extruder is used. As is the case throughout this book, the emphasis is on the behaviour and response of the polymer materials, and on description of the principles of the processes.

The blow moulding process, although really a downstream extension of extrusion, constitutes a separate sector of the industry and so merits a chapter to itself, and the same applies to injection moulding: in both these cases there are features of the moulding half of the process which require as much attention as the extrusion section.

5.1 Profile extrusion

5.1.1 *Profiles and dies*

Profile extrusion is the direct manufacture of product from the extruder die. Thus, of necessity, these products are continuous lengths whose cross-sectional profile is determined by the die shape. Examples are

> yellow polyethylene gas piping
> PVC water and drainage pipes
> PVC window frames
> PVC rainwater goods – gutters, downspouts, etc.
> sealing strips for car windows, windscreens, doors, etc.
> curtain rails
> garden hose
> flat and corrugated sheeting for, e.g. roofing.

Design of dies for such processes, as we have seen, is never an easy matter. Figure 5.1 is a schematic drawing of a pipe or tube die. The main problem is to create an annulus through which the pipe will emerge. This is achieved by having a centrally located mandrel, or 'torpedo', held in place by means of the 'spider'. The spider is a device, usually with screw adjusters, which centralizes the mandrel. At once we can see that the mandrel itself, and

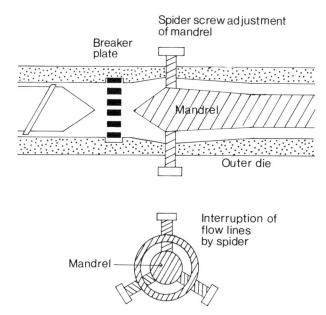

Fig. 5.1 Schematic diagram of a pipe die.

especially the spider legs, will interrupt the melt flow. This disrupts the streamlines and tensional components appear. The gradual tapers, followed by a long land, are devices to maintain and restore laminar flow and parallel streamlines, as far as possible, after the flow lines rejoin beyond the spider legs.

Other profiles present even greater problems. If the required profile is not circular the flow patterns and velocities vary at different places, and the die swell is not the same at all points. This means that the die profile has to 'aim off' to achieve the desired shape. Figure 5.2 shows two simple shapes, the dotted lines being the approximate die profiles to produce them.

The extension of this to very complex profiles, such as PVC window frames, requires great design skill and experience and usually much trial and error. The dies for complex profiles are therefore very expensive and are carefully used and protected from accidental damage.

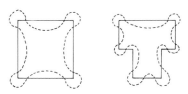

Fig. 5.2 Die profiles: dotted lines show die profiles to produce shapes shown in solid lines.

5.1.2 Downstream operations

Often extrusion of profiles requires, downstream, the addition of a secondary operation to deliver a fully satisfactory product. An example is suggested in Fig. 4.29(b), which shows die swell in an extruded pipe. The die dimensions thus determine the pipe dimensions only approximately, because of the effect of die swell. Usually, some 'drawdown' is applied to pull the extrudate away from the die exit, and this counteracts the swell; the drawdown also helps to orientate the polymer molecules, which enhances linear strength. For some purposes this will suffice, but if more exact pipe dimensions are required, a sizing mandrel is used. The details of this will depend on whether the inside or outside diameter of the pipe is important. If exact i.d. is required, the extruded pipe, while still hot and soft from the extruder, is passed over a mandrel of appropriate size (Fig. 5.3). In other

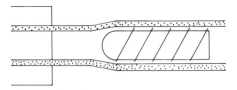

Fig. 5.3 Pipe extrusion: internal sizing mandrel.

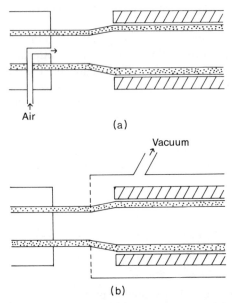

Fig. 5.4 Pipe extrusion – external sizing: (a) pressure sizing; (b) vacuum sizing.

Cross-head extrusion 115

cases the o.d. is specified and here an external sizing device is required. Two basic versions exist: in one the pipe is pressurized against the external mandrel by an air injection, whilst in the other a vacuum outside the mandrel allows the normal atmospheric pressure inside the pipe to hold it against the mandrel (Fig. 5.4). Other downstream operations are the haul-off of the extruded product, also to be seen in Figs 5.3 and 5.4, and reeling or cutting and stacking.

Figure 5.5 shows extrusion of sheet, thick or thin. The design of dies for sheet extrusion and some of the downstream processes which use the sheet are described in Section 5.6. Corrugated sheet is made by extrusion of a flat sheet, which is cooled and cut to size. It is then reheated by IR heaters and deformed to the corrugated shape in a press in tandem with the extrusion line, so that the finished corrugated sheets collect at the end of the production line.

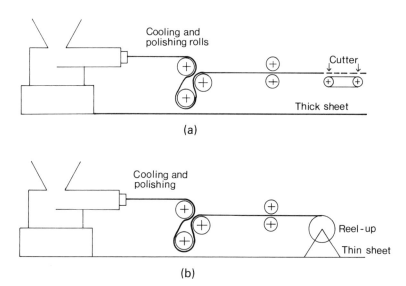

Fig. 5.5 Sheet extrusion – thick and thin.

5.2 Cross-head extrusion

A derivation of the conventional profile extruder is that in which the melt, or rubber, turns through 90° before emerging from its die. There are two principal subdivisions of the industry which use this technique. In the first it is used to cover substrate with an extruded covering and the second is the extrusion of preforms for blow moulded containers. The biggest sector of the industry in the first category is electrical cable manufacture where polymer insulation, often polyethylene, is applied to electrical conductors by this means. Figure 5.6 shows the process schematically. The cable is drawn through the coating head from a reel upstream of the extruder. It enters with an interference fit which prevents any tendency for backwards flow of polymer. It proceeds right through the die and downstream to the haul-off gear of the plant. The molten polymer enters at right angles and surrounds the cable. The coated cable is thus pulled through by the haul-off, the die forming the polymer to the correct diameter.

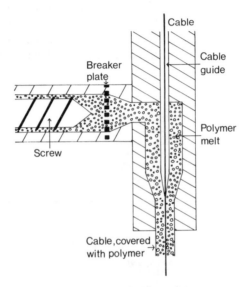

Fig. 5.6 Cable covering by cross-head extrusion.

The other main use of this principle is in the manufacture of fire hose. The substrate in this case is a circularly woven textile 'jacket'. The extrusion requirement is a pipe or tube so that a central mandrel is needed, over which the jacket can pass. In one version of this process the design of the cross-head chamber allows the jacket to be coated simultaneously inside and out to *line* and *cover* the hose in a single operation.

The fire hose process is an example of extrusion of rubber. The lining and cover of the hose is rubber. Details of rubber compounding and technology appear in Chapter 10, and it is sufficient here to say that the rubber compound is unvulcanized at this stage, the extruder running at 100–110 °C. Typical rubber extruders are used, with L/D ratio 5–10 and low compression ratio of about 1.5:1. The rubber is vulcanized later by filling the hose with steam at about 100 p.s.i. (6.5 atmospheres, 170 °C).

5.3 Orientation in pipes and hoses

During discussion of the principles of extruders and dies the topic of the orientation of the polymer molecules has arisen several times. We have seen effects which result from earlier processing history (memory) and others which depend on the elastic component in the deformation of the chains (e.g. die swell). Dies are constructed to have long lands in order to align the molecules. We may thus ask what are the consequences of alignment or orientation of polymer molecules in the finished product? As we have seen (Chapter 1), polymer physical properties – strength, modulus, extensibility – reflect the long-chain structure of the molecules: the relatively low modulus and tensile strength depends on the uncoiling of chains rather than the breaking of bonds. Orientation of the chains should cause these properties to become anisotropic – different in different directions, and this indeed is what is observed. This is because the intermolecular forces become anisotropic under stress, as the random coils are distorted. Holding a stress for a long time allows stable rearrangements to occur – this is the physical basis of viscous flow and creep: rapid or short-term stresses invoke elastic response as the random coils try to recover their preferred arrangement. In the previous chapter we saw how 'Deborah Number' can be used to quantify this.

Orientation is important in most polymer processes and products. In extruded products the emphasis is usually on an alignment in the direction of travel, the *machine direction*. Products made by profile extrusion have markedly different properties in the machine direction and across it. Thus, if an extruded pipe is tested by pressurizing it until it bursts, failure occurs in a slit-like manner along its length. The greater strength is along the orientated chains, and failure will be in the weaker forces holding the chain across the machine direction.

5.4 Orientation and crystallization

As we saw in Chapter 1, some polymers are partially crystalline. Polymers like polyethylene and acetal are well above their T_g and yet are stiff, hard materials at room temperature because they have appreciable crystalline content. The absence of crystalline structure would leave them as rubbers. Atactic polypropylene illustrates this; it is a rubbery material used for backing domestic loose-lay carpet tiles. Orientation in crystalline polymers has an important effect on their properties. This is illustrated by the development of properties in tubular blown film, one of the most important extrusion processes, used in the production of vast quantities of polyethylene film for very many outlets.

5.5 Tubular blown film

5.5.1 *The process*

The extruder is fitted with an annular die, pointing (usually) upwards. The tube produced is inflated with air and at the same time is drawn upwards in a continuous process. The air inside is contained as a large bubble by a pair of collapsing rolls at the top (Fig. 5.7). The polyethylene may be thought of as sliding over a pressurized gaseous internal sizing mandrel. It expands outwards to about three times its original diameter and at the same time is drawn in the other direction. The result is that it becomes *biaxially orientated* and this orientation is made permanent by the crystallization which freezes the orientation in place.

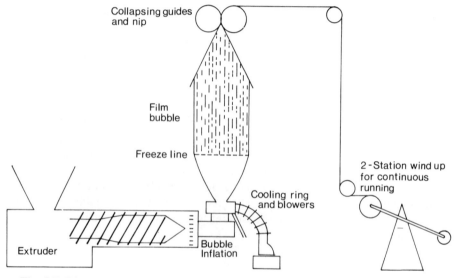

Fig. 5.7 Diagram of blown film process.

It is worth looking at this process in a little more detail. On first acquaintance it looks distinctly odd! Why does the bubble not burst? There must be a stabilizing mechanism. To find it we must return to the overall behaviour of the material, and think about the nature of the process. A moment's thought will lead to the conclusion that the bubble blowing is mainly a tensile rather than a shear process. As we saw in Chapter 2 there are four components to deformation: viscous and elastic responses to both shear and tensile stresses. The main concern in Chapter 2 was with shear viscosity, although tensile viscosity was touched upon. Also, we have seen some of the results of elastic response, mainly in conditions of shear stress. How do polymers react when the dominant stress pattern is tensile?

5.5.2 Response to tensile stress

As for shear stresses, there is a viscoelastic response. Tensile viscosity λ tends to be high (approx. 3η). There is also a parallel to Newtonian and non-Newtonian behaviour. If the tensile viscosity is independent of strain rate it is said to be *Troutonian* (equivalent to Newtonian in shear behaviour). There are two types of strain-dependent, non-Troutonian behaviour: tension stiffening and tension thinning. These are shown in Fig. 5.8.

Most polymer melts are Troutonian, e.g. PMMA, PS, PC, nylon, PET. Branched LDPE is tension stiffening. Linear polyolefins, e.g. HDPE, PP are tension thinning. The elastic response is always tension stiffening.

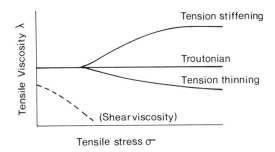

Fig. 5.8 Tensile viscosity.

5.5.3 Stabilization of the bubble

Now we can see how the tensile response stabilizes the film bubble. Such a system will always contain adventitious irregularities. Where the extrudate is slightly thinner the stress is greater, and a non-stiffening material will strain further, possibly to rupture. Tension stiffening, however, leads to increased viscous or elastic response, which more than compensates for the increased stress and the system stabilizes.

The other important stabilizing influence is the crystallization rate. When a film is manufactured in this way there is always a characteristic *freeze line* or *frost line* to be seen a little way up the bubble. This is where the polymer is crystallizing and becoming less transparent. For a polymer that runs well in this process the crystallization is itself tension-induced and hence tension-stiffening; the crystallization rate should not be so high that the biaxial orientation cannot be established first. Polymers in which this happens, e.g. acetal and nylon, usually cannot be processed satisfactorily by this route.

The dominant effects are the elastic response in the cooling, rubbery extrudate and the crystallization rate, both tension stiffening, so that even HDPE whose melt is tension thinning, is overall tension stiffening. LDPE was one of the first polymers in which the process was applied, and its tension stiffening viscosity was important in the early development of the process.

An interesting contrast exists in the production of polypropylene film. Polypropylene is tension thinning in the melt and also its crystallization rate during cooling is rather slow; this makes it difficult to use in the conventional film blowing process described above. A different technique is adopted in which the extrudate is quenched from the melt with ice water to give a rubbery amorphous tube; this is then reheated to the temperature at which crystallization is at its maximum and then blown (Fig. 5.9). Note that the polypropylene process is run vertically downwards. Blowing the reheated tube avoids the problems associated with a tension thinning melt and a slow crystallization rate which would result in an unstable bubble. This process for polypropylene is particularly interesting because it foreshadows the principle of stretch-blowing of bottles, which has gained prominence in recent years for packaging carbonated drinks.

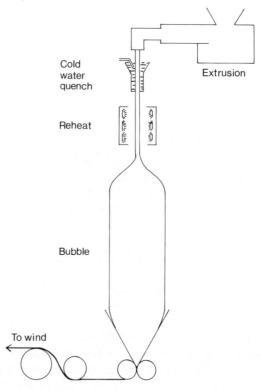

Fig. 5.9 Polypropylene film blowing.

Other film and sheet processes 121

Blown polypropylene film finds extensive uses in packaging. It is the 'crackly' film that will not screw up for disposal, often used for potato crisp packets and on the outside of tea and tobacco packets. The tobacco film needs to be particularly impermeable to gases, to retain moisture levels and aroma in the contents. To improve its properties in this respect it is coated with poly vinylidene chloride from an aqueous dispersion and dried.

5.6 Other film and sheet processes

5.6.1 *Film and sheet dies*

Film and sheet can also be produced as direct profiles; thicknesses below about 0.5 mm are usually called film and above this gauge are referred to as sheet. Also, there are several downstream processes which use directly extruded film or sheet.

The design of dies for the production of sheet or film presents certain problems. The requirement is to deform the melt from its essentially cylindrical shape as it enters the die zone to a wide, thin form, and to maintain even heat-history, temperature and pressure profiles whilst doing so. Three basic die forms are used; these are shown in Fig. 5.10.

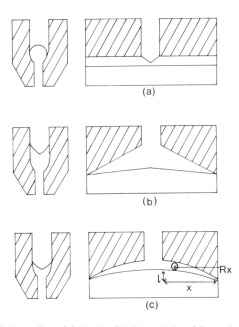

Fig. 5.10 Film and sheet dies: (a) T-die; (b) fishtail die; (c) coathanger die.

The T-die is obviously the simplest and cheapest to manufacture but it does not give good distribution of melt, nor is it well streamlined. It is not really suitable for high-viscosity or easily degradable melts. This design is used for extrusion coating (see below) where low molecular weight and hence easy-processing polymers are often suitable. The fishtail design gives better melt distribution; however, especially for wide or thick sheets, it is massive and contains a large mass of polymer which can entail temperature and degradation problems. The most commonly used geometry is the coathanger die (Fig. 5.10(c)). This is the most difficult to machine, and hence the most expensive, but it gives easily the most uniform melt distribution. The feature of it is the coathanger-shaped, circular cross-section distribution channel which feeds the land. The geometry can be described by remarkably simple expressions relating radius and distance [1]

$$R_x = R_o [x/b]^{1/3}$$
$$\text{(land length) } L_x = L_o [x/b]^{2/3}.$$

Dies for sheet and film are usually fitted with a *flex lip* which allows the sheet thickness to be adjusted by means of a series of bolts across the sheet width. Often, there is also provision for adjustment to the distribution channel by means of a similarly bolt-operated *choker bar*. This allows for compensation for changes in polymer type or output conditions, both of which can influence the power law index of the melt and hence the distribution characteristics.

5.6.2 Cast film and extrusion coating

In the production of *cast film* the film is extruded directly on to rollers which control its cooling and maintain its flatness. The rollers can also impart some degree of drawdown to enhance the film properties by orientation; the thickness of the film decreases if drawdown is used. Usually, the die for cast film is deflected downwards to effect the best approach to the casting rolls.

The process for *extrusion coating* is similar. The film emerging from the die contacts the substrate to be coated at a nip, and the resulting laminate

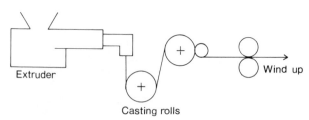

Fig. 5.11 Cast film process.

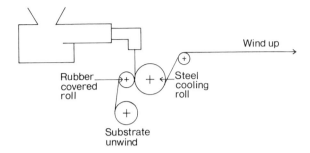

Fig. 5.12 Extrusion coating.

proceeds round a cooling train to the wind-up. Cardboard coated in this way is extensively used for foodstuffs, e.g. milk and soft drinks.

These processes are illustrated in Figs 5.11 and 5.12. The extruders are shown with a cross-head die, for clarity, but conventional in-line dies are just as easily used.

5.7 Synthetic fibres

A further derivative of profile extrusion is the *spinning* of synthetic fibres and monofilaments for use in textiles and industrial applications. Polymers extruded from the melt to produce fibres include nylon, and PET (e.g. *Terylene, Trevira*) and polypropylene. Many other polymers are used for fibres but may not be made from melts: some are spun from solution.

The dies are usually *spinarets* which are multi-hole dies. The emerging bundle of fibres is quenched in water or a countercurrent of air, under a high drawdown rate. The fibres are usually then further highly cold-drawn as they pass to the wind-up, to orientate the molecules and so develop the necessary linear strength.

Just prior to the spinaret there is often a fine filter of, e.g. sand. This becomes progressively blinded as extrusion proceeds, which increases the pressure drop across it, eventually to the point at which the fibre integrity cannot be maintained. To overcome this the extruder is frequently used to supply a gear pump, which then delivers a constant pressure feed to the spinaret by virtue of its positive pumping action. A curious fault that sometimes affects fibre spinning is *draw resonance*. Regularly spaced 'blobs' appear in the highly drawn fibre. The cause of this can be traced back to the tensile stress to which the fibre is being subjected. This tension can detach the extrudate from the die wall in the same way that 'sharkskin' occurs. The detachment travels backwards down the die until a 'blob' of melt from the pre-die mass is pulled directly through the die in a surge: the extrudate is momentarily adhered to the die lip again, but then the detachment starts again under the influence of the high draw, until it once more reaches the die entry, and pulls another blob of melt forward.

5.8 Netting

Netting products for purposes as diverse as garden uses, fruit and vegetable packaging and land reinforcement are made with annular dies, as for tube, but with the outer die and the central mandrel counter-rotating and close fitting. Both parts have grooves or slots so that concentric sets of filaments extrude; when counter-rotation starts the filaments cross one another to form welded junctions and a net pattern. Downstream biaxial orientation by drawing over a larger mandrel orientates and strengthens the net.

This process was invented and developed by the Netlon company, who later developed the orientation aspect with their '*Tensar*' product. In this a flat extruded sheet is punched with a regular pattern of holes. It is then subjected to a high degree of biaxial orientation, which draws the molecules in both directions. The original circular holes become the spaces in a net-like structure and the strands lengthen and develop great strength through orientation. This product finds application in heavy duty land reinforcement.

5.9 Co-extrusion

Co-extrusion is the extrusion of more than one type of polymer at once to give a laminated product. Clearly, this requires a separate extruder for each polymer, the multilayer product forming at the die. The technique allows products to have different properties on each side, or more usually, inside and out. Thus, an inner layer might confer impermeability, sandwiched between outer ones which would have superior abrasion resistance. Often, it is necessary to provide *tie-layers* which bond the functional layers together. Thus, the sandwich would actually have five layers: outer–tie–central–tie–outer.

In a simpler, familiar example two layers are used. The inner containers for many brands of breakfast cereals are made from blown HDPE film with an inner layer of lower softening temperature; this allows sealing of the packet in a heated crimping press, when the inner softens and seals without damaging the outer. In the early days of this packaging system there was difficulty in getting the processing conditions correct, and the packets either did not seal properly or were sealed so tightly that they could not be readily opened! However, the technique is now well controlled, with better selection of polymer grades for the two layers, so that these sealing problems are overcome.

As indicated above, co-extrusion techniques are employed not only for profiles but also in blown films and, as we shall see, in some examples of blown bottles. Why do the polymer streams not mix when they come together at the die? The reason is that the conditions for laminar flow prevail, rather than those for turbulence which would lead to mixing. The high viscosity is the main contributor with the result of a low value for the Reynolds Number, Re

$$Re = \frac{\text{density} \times \text{velocity} \times \text{channel dimension}}{\text{viscosity}}.$$

Under these conditions the Reynolds Number is very low (< 10) because of the high value for viscosity: a value in excess of 2000 is usually regarded as the threshold for the onset of turbulence.

Reference

1. Rauwendaal, C. (1986) *Polymer Extrusion*. Hanser, New York, p. 443.

6
Blow moulding

6.1 Blow moulding principles

Blow moulding is the established technique for producing bottles and other containers based on simple hollow shapes. There are two major subdivisions – extrusion blow moulding and injection blow moulding. Extrusion blow moulding was formerly by far the dominant technique but in recent years injection blow moulding has emerged as a major method for the production of bottles for carbonated drinks, especially using poly(ethylene terephthalate) (PET).

The principle of blow moulding has been used for centuries by glass blowers. A semi-molten tube is formed; this is clamped between the two halves of a split mould and is inflated with air to fill the mould. The mould surfaces are cooled so that the product is frozen into shape whilst still under air pressure. The product is then recovered by opening the mould.

In extrusion blow moulding the semi-molten tube, called a *parison*, is formed directly from the extruder, ready hot and soft. In injection blow moulding the tube, here more usually termed the *preform*, is made by injection moulding and is reheated to blowing temperature. These two variants are discussed separately below.

6.2 Extrusion blow moulding

6.2.1 *Extrusion*

The principle of the process is shown in Figs 6.1 and 6.2. The extrusion process may be continuous, when it will be necessary either for the parison to be cut off and moved to mould or for the mould to move away carrying the parison. Alternatively the extrusion can be intermittent, when the mould can remain beneath the extrusion point. The former arrangement is the more usual because it allows faster production rates.

Figure 6.1 shows the most common arrangement, with downward extrusion. Two important effects result from this: 'parison' sag, caused by gravity acting on the semi-molten extrudate, and die swell. These effects tend to oppose one another to some extent, but together they act to give

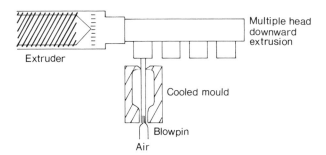

Fig. 6.1 Typical extrusion blow moulding arrangement.

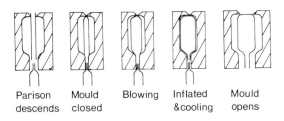

Fig. 6.2 Stages in blowing a bottle.

parisons with thick-walled bottoms and thin-walled tops: early in the parison extrusion die-swell thickens the walls; later, the increasing weight stretches the parison to thin the walls. The device shown in Fig. 6.3 is used to overcome these effects; it is a variable mandrel called the *parison variator* and it varies the wall thickness through the extrusion.

Let us look in a little more detail at the material response during the extrusion part of the process. As it descends the parison is clearly under tensile rather than shear forces. The process time is of the order of 1–5 s and the relaxation or natural time of most polymers is longer than this under these conditions. The process is thus mainly elastic in nature. The tensile elastic modulus is $\simeq 10^4 \, \text{Nm}^{-2}$ and the tensile stress is 10^3–$10^4 \, \text{Nm}^{-2}$, so that quite large deformations are involved. The parison variator is, however, easily able to accommodate such changes.

Figure 6.2 shows a 'bottom blow' arrangement; the parison descends on to the *blow pin* (also known as the *spigot* or *blowing mandrel*). The advantage of this method is that there is no delay between mould closure and blowing: its disadvantage is that a scarred bottle neck is often obtained because the parison has to be large enough to descend over the pin, and extra trimming after moulding is needed. For a 'flash-free' bottle neck, perhaps designed for a screw or other sealing cap, 'top blowing' can be used; however this requires time for insertion of the pin.

128   Blow moulding

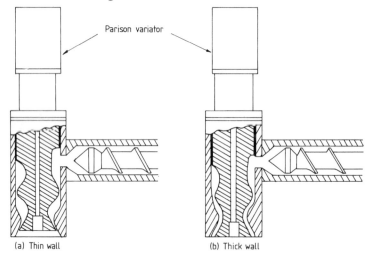

Fig. 6.3 Parison variator.

In either case, the bottom of the bottle is formed by the 'pinch-off' by the mould, and a characteristic of blow moulded plastics bottles is the scar caused by this mould closure weld. The action of the pinch-off has always been a source of trouble for bottle blowers. Sharp cutters would lead to a neat, well formed weld but they work best with slow movement of the mould halves, because of the elastic response of the semi-molten polymer at high rates. However, good moulding and rapid output rates demand fast mould closure, which best suits blunt cutters, and these give poorly separated welds and unsightly scars. The combined effects of the pinch with reasonably sharp cutters and rapid bottle expansion tends to give thin welds, which are weak. Over the years various devices have been introduced to squeeze extra material into welds, but this remains something of a 'black art'.

6.2.2 Blowing

The hollow product, as we have seen, is made from the parison by its expansion with air. The similarity to film blowing is at once apparent, and the evenness of wall thickness of the blown product depends on the same stabilizing effects. Once again, the elastic response is important. If we recall the concept of Deborah Number

$$N_{deb} = \frac{\text{relaxation time of material under prevailing conditions}}{\text{timescale of process}}.$$

For the process to be predominantly elastic, and hence stabilizing by tension stiffening, N_{deb} should be > 1. This could be achieved by using

- a lower temperature, which would lengthen relaxation time
- reducing the process timescale, i.e. blow rapidly.

Usually, the first option is not practicable and so a rapid blow is adopted. However, too rapid a blow may cause rupture to occur, if the tensile strength of the soft parison material is exceeded (see the calculation below), or may entrap air bubbles between mould surface and moulding to leave blemishes on the product. As usual in manufacturing processes, some compromise is needed to give optimum results, and this compromise is usually established by experience on the particular plant. It is a fairly straightforward matter to calculate the wall thickness of a blown bottle if the dimensions of the die and the amount of die swell are known [1]. Figure 6.4 shows a cross-section of a

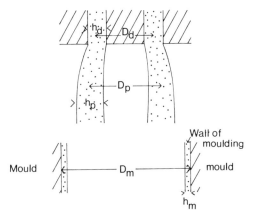

Fig. 6.4 Wall thickness of a blow moulding.

parison emerging from a tubular die: it will be blown to form a bottle, and we shall derive an expression to find the wall thickness of the bottle. We can also find the maximum allowable blowing pressure to avoid rupture of the parison. In Fig. 6.4

D_d is the die diameter (mean)
D_p is the parison diameter
D_m is the mould diameter
h_d is the die annulus width
h_p is the parison thickness
h_m is the moulding wall thickness.

We wish to find h_m.

Swelling of thickness of parison, $B_t = \dfrac{h_p}{h_d}$.

Swelling of parison diameter, $B_p = \dfrac{D_p}{D_d}$, i.e. $D_p = B_p D_d$.

130 Blow moulding

It can be shown that

$$B_t = B_p^2$$

Therefore

$$\frac{h_p}{h_d} = \frac{D_p^2}{D_d} \text{ or } h_p = h_d(B_p)^2.$$

Inflate to mould diameter D_m, and assume constant material volume and no draw down

$$\pi D_p h_p = \pi D_m h_m$$

$$h_m = \frac{D_p h_p}{D_m} = \frac{D_p(h_d B_p^2)}{D_m} = \frac{B_p D_d(h_d B_p^2)}{D_m}$$

$$h_m = \frac{B_p^3 h_d D_d}{D_m}.$$

Thus, knowing die dimensions, mould diameter, and diametric swell ratio, we can find the wall thickness of the moulding. B_p can be found by direct measurement, or from B_t and $B_p = \sqrt{B_t}$ if the swell of the wall thickness is more easily measured.

As an example we will find the wall thickness of a blow moulded container made using a parison die of i.d. 40 mm and o.d. 44 mm. There is a parison wall thickness die swell ratio of 2.3. The container mould has a diameter of 100 mm.

$$h_d = \tfrac{1}{2}(44 - 40) = 2 \text{ mm}$$
$$B_p = \sqrt{2.3} = 1.517$$
$$D_d = \tfrac{1}{2}(44 + 40) = 42 \text{ mm}$$
$$h_m = (1.517)^3 \times 2 \times \frac{42}{100}$$
$$= 2.93 \text{ mm}.$$

We can also find the maximum permissible pressure to avoid rupture of the parison, using the following data.

Tensile strength of semi-molten parison = 10^7 Nm^{-2}.

The Barlow formula which relates hoop stress, dimensions and internal pressure for a pipe is:

$$\sigma = \frac{PD}{2h}$$

$$P = \frac{\sigma 2h}{D}$$

where σ is hoop stress,
P is internal pressure,
D is diameter of pipe and
h is wall thickness.

In the example above,
$h = h_m = 2.93$ mm and
$D = D_m = 100$ mm.

Maximum allowable stress is the tensile strength of the material

$$= 10^7 \text{ Nm}^{-2} = 10 \text{ MPa}$$

$$\text{(maximum pressure) } P = \frac{10 \times 2 \times 2.93}{100} = 0.59 \text{ MPa}.$$

6.2.3 Product properties

As we shall see in more detail in Chapter 8, one of the effects of the large coefficients of thermal expansion exhibited by polymers is a large shrinkage as mouldings cool, and this effect is exaggerated by density changes as semi-crystalline polymers crystallize. Economic considerations will demand rapid cycle times, with ejection from the mould at the earliest possible time and highest temperature, but this will tend towards some shrinkage in the product. Conversely, a longer dwell time in the mould to allow more complete cooling and perhaps a better surface and dimensions in the product increase production costs. No hard and fast rules can be stated about the balance between these conflicting factors; processes have to be tuned to individual product and market requirements.

The best surface appearance, usually a glossy finish, is obtained from a slightly roughened mould surface. The mould interior is given a sandblasted surface. Thus the actual mould surface is not reproduced; the gloss comes from the original extrusion expanded by the air. There are other examples in polymer processing of glossy finishes from cold dull moulding or casting surfaces: glossy surfaces can be made on PVC leathercloths using dull, plain, cold embossing rollers. These surfaces form as a molten glaze; they freeze immediately upon striking the cold mould or embossing roller. If a matt finish is required it is usually necessary to heat the moulding surface to allow time for the polymer to flow before solidifying.

An important property in a blown bottle is its ridigity, which depends on the stiffness of the walls. The stiffness depends on the thickness of the walls; the stiffness of any component is related to thickness by a cube law. It also depends on the flexural modulus of the material, which for polymers is

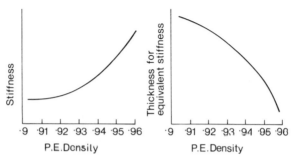

Fig. 6.5 Density–thickness–stiffness relationships in polyethylene.

related to the degree of crystallinity: the density of the polymer also increases with crystallinity.

The stiffness and density of polyethylene bottles will vary not only between PE grades but also within a single batch if processing conditions vary. Bottles packed slightly warm into boxes will be stiffer than those demoulded warm and air cooled before packing, and these in turn will be stiffer than those fully cooled in the mould, especially if refrigerated cooling

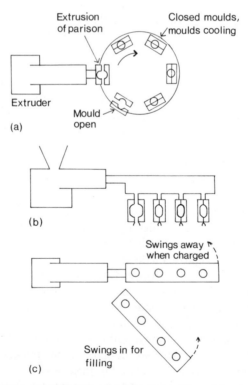

Fig. 6.6 High output arrangements for blow moulding: (a) carousel of moulds; (b) multiple moulding station; (c) alternating sets of mould assemblies.

Injection blow moulding 133

water is used. The reason, of course, is that slower cooling allows more and larger crystals to form. Demoulding and packing slightly warm might thus be the most economical method not only because this shortens the cycle time but also because a marginally thinner bottle can be made to achieve a given stiffness.

Similarly, it can sometimes be more economical to mould bottles from HDPE, even though it is a more expensive polymer than LDPE, because its greater crystallinity allows thinner bottles to achieve the required stiffness, this saving outweighing the cost-per-kg disadvantage. Whether this is so will depend on the current relative costs of the two polymers. These effects are illustrated in Fig. 6.5.

Bottles form the largest product group made by blow moulding, but there are other related products. These include large blow mouldings such as drums for chemicals, including speciality products for acids, coolant expansion tanks, traffic light housings, petrol tanks. It is thus a general method for the production of seamless hollow goods. Such goods cannot be made by injection moulding because they cannot be removed from the mould unless they are cylindrical, i.e. no narrow neck. In Fig. 6.6. some ways in which high outputs are achieved are shown diagrammatically.

6.3 Injection blow moulding

6.3.1 *Background*

Injection blow moulding has come into prominence in recent years for the production of bottles for carbonated drinks. It differs from the extrusion process described above by using an injection moulded preform instead of a directly extruded parison. The preform is moulded into a very cold mould, often employing refrigerated coolant, to quench it in its amorphous state. The preform is reheated to just above its glass transition and *stretch blown*. The stretch blowing is accomplished by pushing down the blow pin, which stretches the preform downwards, and simultaneously blowing to give radial expansion. Once again we see a process which develops biaxial orientation in the product. The alternative name for this process is *stretch-blow*.

A clear parallel can be drawn between this process and that used for polypropylene film blowing (Section 5.5.3), and in fact polypropylene is one of the polymers for which the stretch-blow process is used. Extrusion blow moulding, by contrast shares the principles of film blowing from the molten polymer, as used for polyethylene.

The polymer most widely used in injection blow moulding is poly(ethylene terephthalate), PET. This is the material found in the millions of bottles for carbonated drinks manufactured today. This new type of bottle emerged as a consequence of the desire to sell Cola drinks in larger packages. The Cola manufacturers found increased consumption when the

134   Blow moulding

product was supplied in larger packages, but that the potential gain was likely to be lost if glass bottles were used because of increased weight and breakages. The final impetus was the 1973 oil crisis which sharply increased the cost of the energy required to produce glass and made plastics bottles attractive economically.

It is interesting to observe that increases in oil prices have a more serious effect on materials like glass and aluminium, which require large amounts of energy to produce them, than on polymers which derive from oil feedstock. The 'total oil content' of glass and aluminium is higher than that of most polymers simply because of the energy used in their manufacture: for glass the energy is required for the high temperatures, and for aluminium the electrolytic process used in extraction from its ore and refining is the large consumer of energy. A number of polymers were tried for the new type of bottle, including PVC, SAN and PET, and PET emerged as the front-runner. Initially, the PET bottle was only attractive economically in a 2 litre size, but improved production methods, and customer usage have allowed smaller sizes to emerge. Usage has spread to many other types of drink, although special coatings are necessary to preserve the carbonation in the less heavily carbonated alcoholic drinks like beer and cider. The most widely used material for reducing the permeability further is poly vinylidene chloride.

6.3.2 *Specification for the new bottle*

The requirements for the new bottle derive from the original product it was to contain, in the first instance, *Coca Cola*, which is heavily carbonated. It contains 4 volumes of carbon dioxide per volume of liquid. The pressure in the head space above the liquid can exceed 5 atmospheres in high temperatures, e.g. in a car interior or boot in the sun. The main requirement, then, is to withstand this pressure

- without loss of 'fizz' in the contents
- without breakage
- without change of shape

A typical test requirement is that after 120 days at 23 °C there should be

- $< 15\%$ loss of CO_2
- no off-taste
- no dilation of shape
- no fall in liquid level (small limits allowed)
- the full bottle should survive a drop test from 2 m height.

The bottle is really a pressure vessel and the best shape for this is a sphere if the material is to be used most efficiently. Obviously, a sphere is not a practical shape and the next best is a cylinder with hemispherical ends. If the bottle is made to an appropriate thickness in its cylindrical section, but with the conventional flat or inverted bottom, the bottom would change shape under the filled pressure unless it were thicker, which would increase its cost. The best practical design is the now familiar one of a cylinder with a hemispherical bottom and a rounded top approximating to a hemisphere but including a threaded neck. A separate base cup is added to enable the bottle to stand up. For this general shape with a volume V and wall thickness of the bottle x

$x \propto 1/V^{1/3}$ for CO_2 loss
$x \propto V^{1/3}$ to limit creep.

Figure 6.7 shows that the permeation limit curve and the creep limit curves cross. At lower volumes the wall thickness is limited by acceptable permeability, whereas at high volumes creep is the limiting factor.

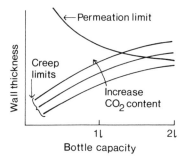

Fig. 6.7 Wall thickness limitations in stretch-blow bottles.

6.3.3 Manufacturing aspects

As we have already seen in outline, these bottles are made by blowing injection moulded preforms, after reheating. During the injection moulding stage the injection speed is limited by control of injection pressure to prevent the formation of spherulitic crystals in the polymer, by stress-induced crystallization. Spherulitic crystals formed at this stage would cause unacceptable haze in the finished bottle. It is also important to control the melt temperature at injection moulding to ensure all crystalline domains have melted, but to avoid the formation of acetaldehyde by degradation of the polymer at higher temperature; about 250 °C is correct. Acetaldehyde causes off-taste in drinks at extremely low concentrations.

The preform wall thickness is limited to 4.2 mm by the cooling rate at injection moulding and the reheat rate before blowing. This limits the bottle

136 Blow moulding

neck size and injection blow moulded bottles (with a few special exceptions) all have the same neck size.

The preform is quench cooled to retain the amorphous state. The reheating is by infrared heaters to above T_g (about 95 °C). The stretch-blow ratio is 3.5 × 3.5, i.e. × 10 overall. This gives wall thicknesses in the region of 0.4 mm.

6.4 Why PET and why stretch-blow?

The problem with polymers for carbonated drinks is to contain the carbon dioxide pressure, and this depends on the gas permeability of the polymer. Among the polymers commonly used for bottles, PET is relatively impermeable (Table 6.1). However, to make a satisfactory article from

Table 6.1 Relative permeability to gases of various polymers

Polymer	Relative permeability
PET	1
PVC	2
HDPE	52
PP (orientated)	57
LDPE	114

PET, or any crystallizable polymer, requires the crystalline structure to be realized. This is the function of the stretch-blow method of making bottles in PET. The same principles can be utilized in, e.g. PVC or polypropylene. PET is a prime example of a polymer whose crystallinity can be controlled by processing: some other crystalline polymers, e.g. acetal, nylon, crystallize spontaneously and so cannot be processed in this way. It is worth looking in a little more detail at the variability of PET in this respect.

If we start with a melt of PET in the range 250–280 °C and quench it the amorphous solid is obtained. This has a T_g of 80 °C and begins to soften above this temperature. If the melt is slowly cooled, large spherulitic crystals form, giving a hard, opaque substance with a crystalline melting point, T_m of 265 °C. If the amorphous solid is reheated to above its T_g (95–100 °C) and then is stretched, stress-induced lamellar crystals form. These are small and the product is transparent. The material is now much tougher and stronger than either the amorphous or the spherulitic crystalline forms. If the orientated crystalline product is now heated further, to about 150 °C, the degree of crystallinity is enhanced and the physical properties improve

Table 6.2 Effects of varying crystallinity in PET

Process		Tensile strength (MPa)
Quench	Amorphous, T_g 80 °C	55
Cool slowly	Spherulites, brittle, T_m 250 °C	
Stretch		
Linearly	Fibres ⎫ 25% crystalline	170
Biaxially	Film, bottles ⎭	
Heat set	Fibres, film 40% crystalline	350

further, and increased temperature tolerance is gained. This is the *heat setting* stage of processing, used for fibres and film. These processing stages available in PET are summarized in Table 6.2. Bottles which are not heat set, are stable up to about 60 °C.

Heat set film is used for 'boil-in-the-bag' food and is stable at 100 °C. It is important to realize that crystallizable polymers like PET and PP, when used in the stretch-blow process, must be quenched to the amorphous solid and then reheated to above T_g to allow stress-induced crystallization to occur. Non-crystallizing polymers may be stretch-blown, and the orientation of polymer chains will confer some improvement in properties, but they can be processed from the cooling melt. If this is attempted with PET, say by simple cooling to 100 °C, spherulitic nuclei will form during cooling and the attempt to stretch-blow will not develop full stress-induced crystalline properties. Also, the spherulitic crystals are large, and this will cause the product to be less transparent, which is unacceptable in bottles and film.

Reference

1. Crawford, R.J. (1981) *Plastics Engineering*. Pergamon, Oxford, p. 63.

7
Thermoforming

7.1 Principles

In thermoforming, a preform, usually an extruded sheet of polymer, is heated until soft and deformed by a shaping force into a mould, where it cools. This is another process where tensile, or elongational, behaviour is dominant.

7.2 Vacuum forming [1]

The most widespread technique for deforming the hot, softened sheet relies on reduction of pressure on one side to allow atmospheric pressure to deform it from the other side. The sheet is drawn into the mould. The force is limited to that available from atmospheric pressure, usually 10–12 p.s.i. Figures 7.1 and 7.2 show the two principal variants. The female mould uses

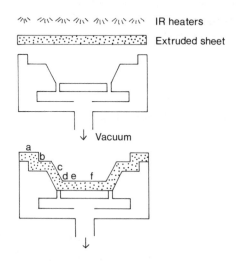

Fig. 7.1 Vacuum moulding with a female mould. Thickness at (a) 1.02 mm (b) 1.02 mm (c) 0.65 mm (d) 0.50 mm (e) 0.50 mm (f) 0.75 mm.

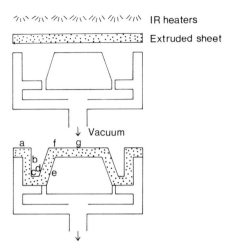

Fig. 7.2 Vacuum moulding with a male mould. Thickness at (a) 1.27 mm (b) 0.65 mm (c) 0.25 mm (d) 0.40 mm (e) 0.50 mm (f) 1.27 mm (g) 1.27 mm.

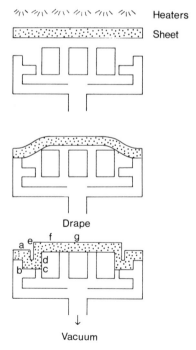

Fig. 7.3 Drape-assisted vacuum moulding. Thickness at (a) 1.02 mm (b) 0.43 mm (c) 0.36 mm (d) 0.60 mm (e) 0.60 mm (f) 0.60 mm (g) 0.60 mm.

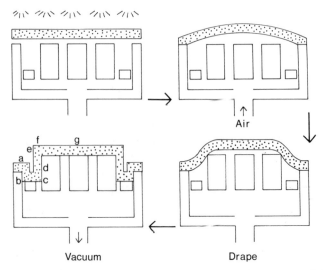

Fig. 7.4 Air-slip assisted drape moulding. Thickness at (a) 1.02 mm (b) 0.43 mm (c) 0.23 mm (d) 0.43 mm (e) 0.60 mm (f) 0.87 mm (g) 0.87 mm.

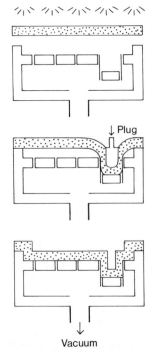

Fig. 7.5 Plug assisted vacuum moulding.

Vacuum forming 141

a cavity for the main formation, whereas the male mould uses a projection. Both figures show a 'large tub' shape being moulded; that in Fig. 7.2 is moulded upside down. The two approaches give differences in the product, the main one being seen in the thickness distribution. The male mould (Fig. 7.2) gives a more substantial base. The female mould (Fig. 7.1) has better thickness at the rims. The male mould gives more waste.

Frequently, the process is improved by using mechanical aids. Some of the principles involved in these are shown in Figs 7.3–7.6. In Fig. 7.3, the hot sheet preforms by draping on the mould first. The vacuum then draws the final shape. This gives a better thickness distribution than simple male moulding, with a greater depth of draw and a smaller blank. The disadvantage with drape assistance is that the preform contacts the cooled mould surface and is chilled, and this reduces the area for stretching. Figure

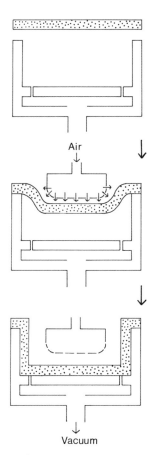

Fig. 7.6 Air-cushioned plug assistance.

7.4 shows how an air-slip technique overcomes this by giving an air cushion to prevent contact chilling. Air-slip gives an even better thickness distribution and a smaller blank. Figure 7.5 shows an item with a deep drawn section to it; a plug is used as a mechanical aid. Thickness control depends on slippage of the sheet on the plug before application of the vacuum, and air-slip could be used to help this. A plug gives good thickness distribution on the deep drawn section, and it is effective in avoiding 'webbing'. This fault occurs when the sheet bridges corners instead of deforming snugly into them. The development of this is seen in Fig. 7.6. A deep drawn article is made in a female mould using an air cushioned plug.

1. The heated sheet sags;
2. Positive air blows it into a bubble;
3. A heated plug blows it into the cavity, with temperature-controlled air jets to prevent direct contact;
4. Vacuum is applied.

These techniques give some idea of the variations possible. They are used in combination to develop processes for individual needs. The figures are redrawn after those in the ICI publication in ref. 1.

7.3 Material stress and orientation

Deformation of the sheet requires work to be done on it while soft but not molten. In this respect, the process resembles extrusion blow moulding. Sufficient strength must be retained to maintain sheet coherence; if the rupture stress is exceeded a hole forms and the process stops.

Considerable biaxial orientation occurs and this, as we would expect, confers good properties on the product: it is also the source of inherent instability, in that vacuum-formed shapes are 'reversible' as we shall see below.

The ideal pattern of deformation would be as shown in Fig. 7.7. This shows a moulding with even thickness. To achieve this would require a material showing easy extension to the required strain, followed by rapid

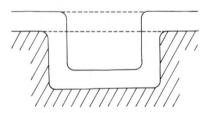

Fig. 7.7 Ideal deformation pattern.

stiffening to stabilize the flow. However, we know that real polymeric materials are not like this. The process is an elongational, free surface flow, similar to blow moulding and film blowing. The best thickness control will result from tension stiffening conditions, and these will be when an elastic response occurs. Elastic behaviour comes when the material is deformed rapidly and so we find the best results, as with blow moulding, when the deforming force is applied rapidly. In vacuum forming this means rapid application of the vacuum; thus we find vacuum forming plant is equipped with a large vacuum reservoir with a capacity of at least four times the volume of air to be removed from the mould cavity.

The importance of tension stiffening is that as the thickness at any point diminishes the stress in that region increases; this causes stiffening which in turn lowers flow, so that the main flow moves to an area of lower stress. The overall effect is towards more even thickness. Material tends to feed from the sides, where expansion, and hence tension, is lower (approximately 2:1 draw), into the corners, where expansion and tension are higher (approximately 5:1 draw).

The inherent instability results from the rapid, elastic deformation. If we think again of the Deborah Number concept, the process time is much less than the relaxation time. The elastic stress is thus 'frozen-in' and can be released under appropriate conditions. Such conditions can be encountered when the moulding is heated above its T_g, which will permit enough molecular motion to release the frozen-in stress, or perhaps subjected to solvent attack. It then attempts to recover its original shape. The most stable shape is usually that acquired at the highest temperature – extrusion as a sheet. An example of this kind of behaviour can often be encountered with vacuum formed polystyrene cups sometimes found in vending machines dispensing hot drinks. These cups are quite satisfactory when used thus; however, if one is used to make a drink (outside the machine) using boiling water from a kettle, the T_g of polystyrene is approached very closely, and the cup will deform. The deformation will be most towards the rim, where the stabilizing orientation is at its minimum.

7.4 Applications

Vacuum moulding is now a well established technique, and three main classes of product may be identified. These are summarized below.

7.4.1 *Thin-wall containers*

These are such items as cups, packs, etc. The moulding line is run in-line with the sheeting extruder. There is accurate control of sheet thickness and extrusion conditions generally to minimize the scrap arisings. This is a very

144   Thermoforming

cost sensitive sector of the industry, working on small margins and large turnover.

7.4.2 *Large technical mouldings*

Here there is a more technical use for the products, and also the scale of the moulding is much larger. The materials (displayed in more detail below) are often more complex. They include, besides simple thermoplastics, composite materials, co-extruded sheets and laminated materials. All must of course be thermoplastic for the process to work. The products include such items as boats, freezer liners, garage doors and domestic baths. There is now no UK manufacturer of enamelled cast iron baths, illustrating the technical and economic efficiency of good plastics products. These products are examples of large area, deep drawn mouldings. The sheeting required is frequently up to 3 m in width, and the extruder die design and control is a key factor for quality control. Many of these products can use as much as 10 m lengths per moulding, and this is thus seen to be large scale plant.

7.4.3 *Skin and blister packaging*

Packaging of this type is used for modern display selling of a vast array of household goods, ranging from hardware items like nails, screws and tools to toiletry items like disposable razors. The so-called 'skin' packs are those where a flexible skin is drawn tightly on to the goods on a rigid cardboard base. 'Blister' packs are preformed foils, roughly conforming to the shape of the article to be packed, sealed to the base after insertion of the article. A variant of the blister is the 'global' pack in which a regularly shaped, often hemispherical and fairly rigid blister is used.

7.5 Materials

The thermoplastics most widely used in thermoforming are

   ABS, Acrylic (PMMA), Polyolefins, HIPS, PVC.

Here we find amorphous polymers coming into their own. ABS and PMMA respond well, the latter being the polymer used for baths. Developments in recent years include the use of co-extruded or laminated sheets for special purposes. An example is the material used for refrigerators. Super-high-impact polystyrene was used in the early days of vacuum formed refrigerator bodies. There were problems with this material from environmental stress cracking (ESC) by milk, fats and oils. There was also some attack on the HIPS by the polyurethane foam insulator. ABS is

Table 7.1 Some examples of materials in vacuum forming

Material	Application	Comments
ABS/S-HIPS	Refrigerator	Ease of forming, oil/fat resistance
Matt ABS/high heat ABS	Motor trade Dashboard area	Non-reflective, good heat distortion
PP/filled PP	Motor trade parcel shelves	Cost effective, rigid
PP UV stable/ PP copolymer	Outdoor applications – garage forecourts	UV stabilizing, expensive – top layer only
Three layer, with scrap in centre	Various	Cost effective

superior in these respects but is a good deal more expensive. The solution was to use a glossy, oil-resistant, but expensive, high acrylonitrile content ABS for the surface co-extruded with cheaper, low acrylonitrile content ABS or HIPS, which also gives easy vacuum forming. Some examples are listed in Table 7.1.

Reference

1. ICI Technical Service Note G 109 (1974) *The Principles of Vacuum Forming*, 4th edn.

8
Injection moulding

8.1 Principles

8.1.1 *The basic process*

The basic principle of injection moulding is to inject molten polymer into a closed, cooled mould, where it solidifies to give the product. The moulding is recovered by opening the mould to release it. An injection moulding machine has two main sections to it:

the injection unit
the clamp unit, or press, which houses the mould.

These functions are described in outline in the following sections, and analysed in more detail later.

8.1.2 *The injection unit*

In the first section the process is virtually the same as the extrusion process already described in Chapter 4. This is the *plasticizing* part of the process. The polymer response is the same, and screw designs, barrel heating, etc., are very similar. The one major difference is that the screw can reciprocate, piston-like, within the barrel during the injection part of the production cycle.

During the plasticizing phase, the output end is sealed by a valve, and the screw accumulates a reservoir, or 'shot' of melt in front of itself by moving backwards against the head pressure. When this phase is complete, the sealing valve opens, the screw stops rotating and pressure is applied to it so that it becomes a ram or piston which forces the accumulated melt through the connecting nozzle into the mould, contained in the clamp unit. This is the *injection* phase of the process.

8.1.3 *The clamp unit*

This is essentially a press, closed by a hydraulic or mechanical toggle system. The clamping force available to it must be great enough to resist the force generated by the melt as it is injected. The pressure in this melt, as shown in Chapter 2, can be around 145 MPa, so that for mouldings with a large projected area the force required can be considerable – in the largest machines, several thousand tons.

8.1.4 *The mould or tool*

The mould is mechanically fastened (e.g. bolted) in the clamp unit, but is interchangeable to allow different products to be moulded. The essential features of a mould are:

1. The *cavity* or *impression*, in which the moulded product forms. A tool may contain a single cavity or several;
2. The *channels*, along which the melt flows as it is injected. These are the *sprue*, which is the channel from the nozzle, and the *runners*, which run from the sprue to the individual cavities. The runner constricts to a narrow gate at the entrance to the cavity. We shall deal in detail with gating in a subsequent section;
3. *Cooling channels*, through which cooling water is pumped to remove the heat of the melt. The size and location of these is often critical in ensuring evenly cooled mouldings;
4. *Ejector pins*, which remove the moulding from the cavity. They are automatically activated as the mould opens.

Figure 8.1 shows the principles of injection moulding diagrammatically.

8.2 The moulding cycle

The sequence of operations for the production of injection mouldings is as follows:

1. The mould is closed. At this stage it is of course empty. A shot of melt is ready in the injection unit;
2. Injection occurs. The valve opens and the screw, acting as a plunger, forces the melt through the nozzle into the mould;
3. The 'hold-on' stage when pressure is maintained during the early stages of cooling to counteract contraction. Once freezing commences the pressure can be released;
4. The valve closes and screw rotation starts. Pressure develops against the closed-off nozzle and the screw moves backwards to accumulate a fresh shot of melt in front;
5. Meanwhile, the moulding in the mould has continued to cool; when ready, the press and the mould open and the moulding is removed;
6. The mould closes again and the cycle repeats.

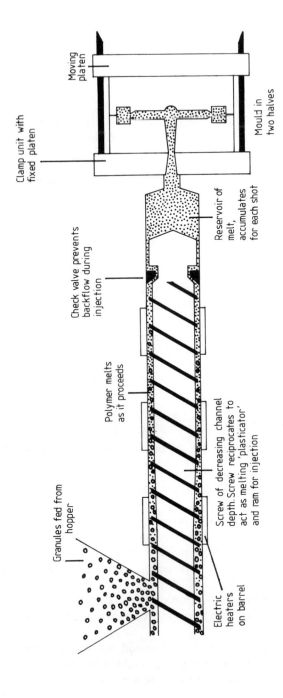

Fig. 8.1 Principles of injection moulding.

If the cycle is displayed in a 'pie diagram' (Fig. 8.2) we can see that a large proportion of the total cycle time is taken by cooling, including, of course, the hold-on time. As a consequence, cooling rates are an important concern in the economics of injection moulding.

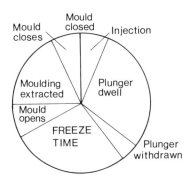

Fig. 8.2 Cycle of operations for the production of injection mouldings.

8.3 The injection moulding machine

The process and essential elements of the machine were reviewed in Section 8.1. In this section the various components are described in more detail.

8.3.1 *The injection unit*

As already noted, the injection unit is essentially a single screw extruder, and the reader is referred to Chapter 4 for an account of its functioning. In essence, we may say at this point that the injection unit comprises an Archimedean screw rotating within a barrel, with minimum clearance between barrel wall and screw flight. The barrel has cuff heaters on it. The screw channel depth decreases from the feed end to the output end to allow compression of the contents. Cold polymer granules are introduced at the feed end and molten polymer emerges from the output end. Heating is partly from the barrel heaters and partly from viscous dissipation as the polymer melt is pumped along by the screw. Unlike a simple single screw extruder the screw in an injection moulding machine reciprocates to effect the injection, in the manner already described. Also, as we have seen, there is a nozzle to connect this unit with the press, and a valve which is closed during injection to prevent backflow of melt past the screw flight, and is open when the screw rotates to allow the fresh shot to accumulate. These valve positions are shown in Fig. 8.3 [1].

150 Injection moulding

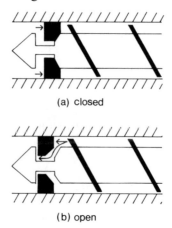

Fig. 8.3 Check valve, open and closed.

8.3.2 *The nozzle*

The nozzle connects the two halves of the machine to allow the melt to pass from the plasticizing stage to the mould. Figure 8.4 [1] shows three variants. In the figure the unshaded areas are the polymer melt pathways. The open nozzle shown in Fig. 8.4(a) is simple and trouble free. However, low

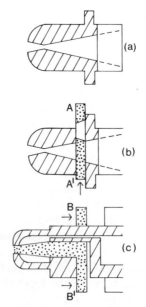

Fig. 8.4 Types of nozzle: (a) open; (b) simple mechanical shut off; (c) needle valve.

viscosity melts like nylon will tend to dribble from such a nozzle, causing contamination and litter, and a positive shut off may then be necessary. In the simple mechanical shut off shown in Fig. 8.4(b), the plate A–A' slides as shown by the arrow. The needle valve shut off in Fig. 8.4(c) relies on injection pressure to open it, with B–B' moving horizontally as shown by the arrows.

Nozzles are usually heated by a heater band, but viscous heating also occurs, because at this point the channel is narrow and thus the shear rate is high. The viscosity therefore decreases and this in turn helps easy injection.

It is usually undesirable for the polymer to freeze in the nozzle after injection and hold-on, and it will tend to do so by contact with the chilled mould, unless prevented. Often, the nozzle will withdraw from contact, or it may be thermally insulated. Frozen polymer in the nozzle has to be remelted at the next injection and results in temperature inconsistency, and hence flow irregularities, in the melt as it proceeds into the mould. This in turn leads to product defects.

8.3.3 *The clamp unit or press*

We have already noted that the function of the clamp unit is to hold the mould closed with sufficient force to resist the injection pressure. This can exceed 140 MPa (20 000 p.s.i.) and 200 MPa may well be needed to prevent 'flash' at the mould mating surfaces. The closure is either by a mechanical toggle or hydraulic.

Clearly, the travel available in the moving half of the press must be sufficient for the depth of the moulding; it must also be sufficient for removal of the moulding, which means more than twice the moulding depth. The closure force needed for a given moulding can be found from its *projected area*. In the example in Fig. 8.5, which represents a tub with a hole in the bottom, like a plant pot, we can see how the projected area includes angled

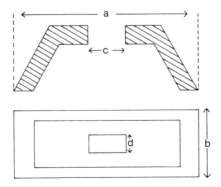

Fig. 8.5 Projected area of a tub moulding.

152 Injection moulding

or radiused side walls but excludes apertures

Projected area = $(a \times b) - (c \times d)$.

The injection pressure applied over the projected area gives the injection force, and hence the clamp force required to resist it. The clamping force available is one way in which machine sizes are categorized; the bigger the

Table 8.1 Sizes of injection moulding machines

Shot capacity (g)	Clamping force (tonnes)
30	10
120	125
350	250
800	375
1500	650
8500	2500

force available, the bigger the machine. An alternative method of size rating is to indicate the maximum shot size the machine can deliver, which is related to the injection unit capacity. Of course, different polymers have different densities; it is usual therefore to relate the shot size to polystyrene. Until recently, shot sizes were expressed in ounces, but metric measure is now used. A rough equivalence between the two rating systems is shown in Table 8.1.

8.3.4 Pressure for injection

The high pressure needed for injection derives from the high viscosity of polymer melts, as shown in Chapter 2. It is applied via the screw, non-rotating, by a hydraulic system. The line pressure of the system is of the order of 7–14 MPa (1000–2000 p.s.i., 70–140 bar). The hydraulic cylinder diameter is 10–15 times that of the screw, which gears the pressure up to that required for injection of the viscous melt.

8.3.5 The mould

As we have seen, the mould is fastened to the clamp plates, at its simplest, in two halves. Figure 8.6 shows such a *two-plate mould*. The figure shows the *sprue* leading from the nozzle to the *runners* which lead on to the *gate* at the entrance to the *cavity* or *impression*. Figure 8.7 shows a *three-plate mould*, which has a third plate between the two on the clamp plates. Three-plate

The injection moulding machine   153

moulds are needed when the runner system and the cavities are in different planes; two openings are required to remove the mouldings and the sprue and runners. There is provision (not shown) to pull the sprue from its channel (a sprue-puller). When the mould opens the solidified polymer fractures at the gate and the moulding is recovered from one daylight; the sprue is pulled (automatically) and falls at the other daylight.

The products shown diagrammatically in Figs 8.6 and 8.7 are similar

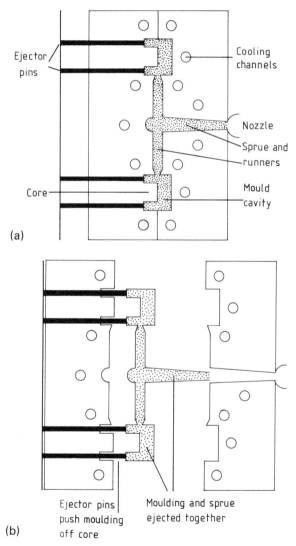

Fig. 8.6 Two-plate mould, closed and opened.

154   Injection moulding

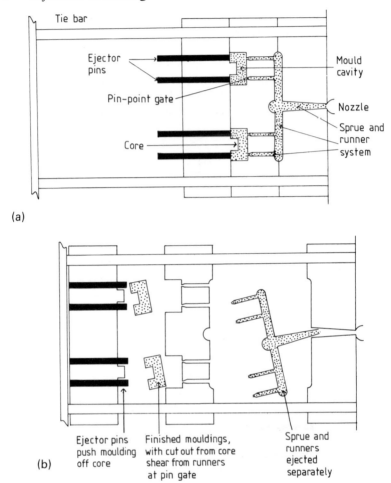

Fig. 8.7 The principle of the three-plate mould.

(a hollow container). The two-plate mould gates from the side and the melt will have to flow round a core to join up on the far side. The three-plate mould gates at two points on the bottom and the melt can flow evenly to form the walls of the moulded vessel. The latter will give a moulding with better properties, with a lower likelihood of distortion.

When the mould fills the air already in it must be vented. Often this happens spontaneously via the ejector pin clearances, but sometimes narrow vents may be provided, of about 0.025 mm diameter, sufficient to vent the air but not to allow melt to enter them. If the venting is inadequate a number of processing or product faults may result. At the most extreme, a bubble of air can be trapped, leaving a dimple in the moulding. A more common fault is 'burning', caused by the rapid escape of venting air; its velocity can be high enough for a temperature rise sufficient to degrade the polymer locally, and to cause characteristic burn marks on the moulding.

This mould is supplied with cooling channels through which water is passed. The temperature of this water varies for different products. Chilled water gives the fastest cycle times but warmer mould temperatures are sometimes required, especially with crystalline polymers, to achieve superior properties in the finished product. The constriction at the gate has three main functions:

1. It allows rapid freezing of the polymer at the conclusion of injection. This isolates the cavity and permits withdrawal of the screw;
2. The narrow and thin solid section allows the moulding to be sheared off easily after demoulding, eliminating finishing trimming in most cases;
3. It increases the shear rate as the melt flows through and hence lowers the viscosity to ease rapid and complete filling of complex shapes.

Several types of gate design are used for different purposes, and some of those commonly used are illustrated in Fig. 8.8. Some of the features of these gating variations are described below.

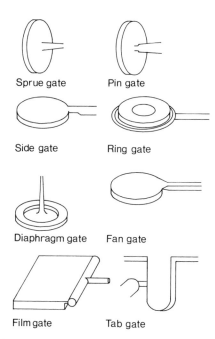

Fig. 8.8 Variations in gate design.

156 Injection moulding

1. Sprue gates. These are the simplest. There is direct feed to a single cavity from the sprue;
2. Pin gates. Pin gates are fed from runners. They are frequently used in three-plate moulds. The small scar they leave leads to easy finishing. The narrow section gives very high shear rate, low viscosity and easy filling in thin mould sections;
3. Side gates. The standard gate type for multi-impression tools. They feed the side of the product. Multi-impression tools should use 'balanced runners' to ensure even pressure distribution through the system. Unbalanced runners can give moulding of unequal quality because the pressure and hence flow is not the same for impressions near the sprue and those remote (Fig. 8.9);
4. Ring gates. These are used for multi-impression moulds making hollow mouldings with the flow round a central core;
5. Diaphragm. Similar to the ring gate but feeding directly from the sprue for single impressions;
6. Fan gates. Fan gates cause the melt to spread fan-wise to cover large areas well;
7. Film gates. Also known as 'edge' or 'flash' gates, these give orientated distribution for flatness in thin flat mouldings. They are much used for transparent products like polycarbonate lensing on meters, where the even flow avoids ripples;
8. Tab gates. The tab eliminates 'jetting' in large plane areas by breaking the flow and making it turbulent as it enters the cavity. Jetting causes unsightly flow lines, especially in transparent materials.

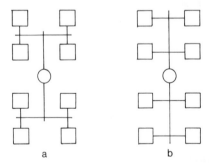

Fig. 8.9 Balanced and unbalanced runners: (a) balanced; (b) unbalanced.

8.4 The polypropylene hinge – a study in gating

This section is developed from the ICI publication *'Propathene for Integral Hinges'* [2]. It provides a study of the effects of different types and positions of gates on the flow patterns of the melt and hence on product quality.

The polypropylene hinge   157

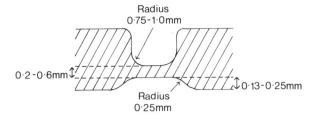

Fig. 8.10 Typical polypropylene hinge configuration.

The polypropylene hinge is shown typically in Fig. 8.10. It is integrally moulded with the rest of the appliance. The most common application for the device is in lidded boxes, e.g. for card index storage, instruments, etc. Other uses include disposable surgical forceps and car accelerator pedals.

The hinge is a thin web of material, usually 0.2–0.6 mm thick, but can be up to 1 mm thick when there is restricted movement, e.g. in the car accelerator pedal. There are radii on either side which minimize restriction during moulding and sharply define the hinge line. The hinge must be straight, or it buckles in use.

The hinge works uniquely in polypropylene, because this polymer has a spiral crystal unit cell, compared with the orthorhombic structure found in polyethylene; other polymers also form rigid crystalline structures. It is important to have highly orientated polymer in the hinge, obtained by even flow through this section during moulding. The spiral structure then becomes highly flexible perpendicular to the orientation direction, rather like the flexibility shown by a thin coil spring; the behaviour of more rigid structures is more analogous to that of rods.

The flexibility is virtually limitless, i.e. the hinge may be flexed repeatedly without fatigue or fracture. On first acquaintance it feels like a fold in a piece of cardboard, which will fail after a relatively small number of flexes; the fact that it does not shows that its structure is quite different from a fold, as described above.

To achieve the required orientation correct gating of the mould is essential. Hesitations or stop–start at the hinge are fatal. Thus, gating close to the hinge is generally unsatisfactory (Fig. 8.11). The melt flows radially from the gate; it meets resistance at the thin hinge section, but not in the rest of the box. It stops at the hinge until the rest is full, then restarts at the hinge.

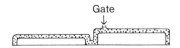

Fig. 8.11 Gating close to the hinge.

158   Injection moulding

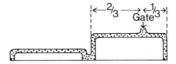

Fig. 8.12 Correct position for a pin gate.

The stop results in a laminated structure in the hinge which delaminates on flexing. Also, in this arrangement, the hinge restriction inhibits filling of the opposite chamber. Gating both cavities is unsatisfactory because a weld line forms at the hinge – even if one is nearer the hinge than the other.

Figure 8.12 shows a pin gate in the orthodox correct position. It ensures full injection pressure behind the melt as it flows through the hinge restriction because the first cavity is already full.

An edge gate on the extreme end of the cavity will also work well. This is

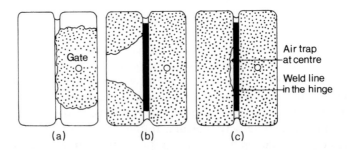

Fig. 8.13 Problems with a single-gated long hinge: (a) the material reaches the hinge but stops while lateral flow occurs; (b) the material stopped in the hinge sets: when flow restarts it is round the edges; (c) the material then flows back towards the hinge from the side of the second cavity, and creates an air trap. This causes either an actual gap or at best a weak mould.

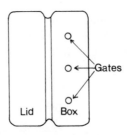

Fig. 8.14 Multiple gating for long hinges.

The polypropylene hinge   159

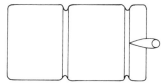

Fig. 8.15 Edge gating.

satisfactory for small items with short hinges; long hinges, as used on boxes, present new difficulties if a single gate is used. Figure 8.13 illustrates these.

Figures 8.14 and 8.15 show how the flow pattern can be corrected. Multiple gates can be adopted, linearly dispersed no further apart than twice the gate–hinge distance or an edge gate is used.

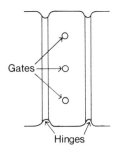

Fig. 8.16 Gating between double hinges.

Figure 8.16 shows how double hinges are made with the gates between them, to avoid a double restriction.

Car accelerator pedals commonly use side gates (Fig. 8.17); the length of such a moulding ensures linear flow through the hinge region. However, if the product is wider than this, diagonal flow could occur, resulting in incorrect orientation; the side gate can then be angled to deflect the melt off the end of the cavity linearly down through the hinge (Fig. 8.18).

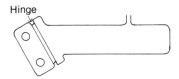

Fig. 8.17 Side-gated accelerator pedal.

160   Injection moulding

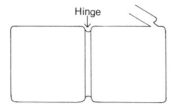

Fig. 8.18 Angled side gate.

8.5 Some aspects of product quality

8.5.1 *Basis of material response*

The injection moulding process can be thought of as a trilogy whose components are

  the injection unit
  the mould
  the polymer.

It remains in this section to deal with the response of the material to its processing and the associated effects on product quality. Various aspects of material behaviour have been described in Chapter 2, which dealt with the physical basis of polymer processing; these included

  heating of the polymer by viscous dissipation and conducted heat
  the viscosities of melts and non-Newtonian flow
  enthalpy and viscosity values for polymers
  non-steady state heat flow during cooling of mouldings.

We have also seen how polymers orientate during processing, and this will clearly impinge on product quality.
  The principal process control parameters are

  temperature of melt
  temperature of mould
  pressure of injection and hold-on pressure
  speed of injection
  timings of the various parts of the process cycle.

However, some difficulties are avoided by good product and tool design in the first place, and we shall therefore first look at some of these, before going on to consider the effects of the process control factors.

Some aspects of product quality

8.5.2 Design aspects

Among the quality problems that can often be minimized by design we may list the following

weld lines
sink marks and voids
stress concentrations at corners, leading to product failure in service
selection of the most suitable material for the product.

(a) Weld lines

These form where polymer flows meet and they can sometimes be avoided; e.g. the tub-shaped moulding shown in Fig. 8.6 will have weld lines, whereas that from the three-plate mould in Fig. 8.7 will not. If welds are unavoidable they can often be moved to a position on the moulding where they are unimportant, by control of the gate position, as we saw in the polypropylene hinge example above. Once design has minimized the incidence, process control can be invoked to minimize the effect. This will entail ensuring adequate temperature and pressure for a good weld. Some examples are given in more detail in ref. 3.

A weld is always a potential air trap, because of the converging melt fronts, and venting may be required at this point in the mould.

(b) Sink marks and voids

These related moulding faults occur when the product section is too thick. The thick part retains heat and is drawn down by contraction forces – especially crystallization which involves a large density change. If the outer skin hardens, and so resists further sinking, internal voids form as the tensile strength of the solidifying melt is exceeded. This is essentially a design problem, to be designed out as far as possible by avoiding thick sections, but it can be helped by careful control of hold-on pressure [3]. When thick sections are required in a moulding, e.g. to confer stiffness, it is better to adopt a modified process such as foam-cored moulding which avoids the problem of sinking and voiding altogether. Alternatively, a pattern of thin-section ribs may serve (Fig. 8.19).

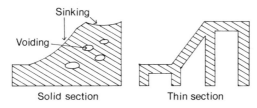

Fig. 8.19 Use of ribs instead of a solid section.

162 Injection moulding

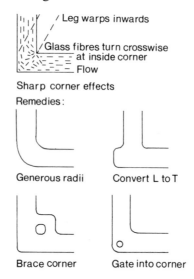

Fig. 8.20 Design features to avoid the effects of sharp corners.

(c) *Stress concentrations*

The consequence of stress concentrations in mouldings with sharp corners can often be failure, especially if the product is load-bearing. Sometimes there will be distortion, especially when fibre-reinforced grades of polymer are used. Figure 8.20 illustrates this and shows a few design features which can assist.

(d) *Computer-aided mould design*

An important development of the past few years has been the emergence of computer-aided design (CAD) methods, pioneered by the Moldflow company [6]. The database of the system contains rheological/temperature/pressure data on polymers of different types. A proposed mould design can be simulated by the computer and the melt flow through it analysed. Different runner and gate sizes and positions can be tried and the optimum found. The technique is particularly valuable for difficult multi-cavity moulds where the flow pattern may be difficult if not impossible to predict. Traditionally such tools were made with undersized channels which were then adjusted by trial and error methods on the plant, a time-consuming and expensive procedure. The Moldflow programme allows the trials to be simulated and a perfect mould to be fabricated directly. The tools for which it has been used include the so-called 'family moulds' in which several different components are made simultaneously. Previously, these tools, though attractive economically because they ease assembly when

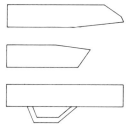

Fig. 8.21 Components in a family mould.

several components of a product can be produced together, have generally been regarded as impracticable. The new technique allows their manufacture with relative ease. An example is a family mould designed for the simultaneous moulding of the three lengths of piping for an industrial vacuum cleaner manufactured by Flymo Ltd. The mouldings are shown diagrammatically in Fig. 8.21. The original design for this mould by the toolmakers [12] had runners arranged as shown in Fig. 8.22(a). It is a conventional symmetrical design, with the runner to the gates and the gates themselves having different sizes, with the shortest component to be moulded at the central position. There was some unease felt that the design might give trouble and the design was subjected to a Moldflow analysis. The runner arrangement to emerge from this is shown in Fig. 8.22(b). It has the

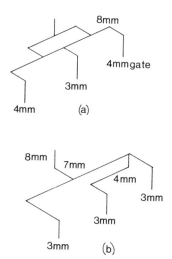

Fig. 8.22 Arrangement of runners in a family mould: (a) conventional design approach; (b) using Moldflow CAD.

164 Injection moulding

same sized runners and gates for all three cavities. This is the best way to ensure even filling of the three cavities, as long as the rate of supply is controlled through the runner sizes. It has long been recognized that this is the ideal, but the calculations to achieve it without CAD techniques, for non-Newtonian, temperature-dependent fluids, are impossibly difficult. The runner layout depicted in Fig. 8.22(b) is rather unexpected, and the designer has stated that it is a layout that would not have occurred to him in the normal way. The new design was made and put into production, where it gave perfect mouldings right from the start.

(e) *Polymer selection*

The subject of selection of the correct polymer for a given application is a very large one. It is not really possible to draw up a comprehensive guide; much depends on individual experience and often several polymers will do the job equally well. In such cases the final choice will depend on cost and the processor's preference. Some examples may be found in ref. 4. However, once again the advent of cheap computer power has made possible a type of CAD approach to this problem. Databases are used which contain the general, mechanical, electrical, etc. properties of a large number of polymers; these are accessed to match the design property requirements in the product, and suitable materials are selected by the computer. An example is the 'EPOS' system, jointly produced by ICI plc and LNP Plastics Ltd.

8.5.3 *Effects of shear heat and pressure*

The pseudoplastic response of most polymer melts under conditions of high shear has already been described in Chapter 2. In the narrow runners of an injection mould the shear rate is about $10^3 \, s^{-1}$, and at the even narrower constriction of the gate it is of the order of $10^5 \, s^{-1}$. The whole system, of course, functions only because the polymer behaves pseudoplastically. It is only because the apparent viscosity is so low at the gate that intricately detailed mould features are faithfully reproduced in the moulding.

Another effect of high shear is the generation of heat – an important aspect of heating the polymer as it moves down the barrel. Thus, during injection, the temperature rises in proportion to the pressure drop involved as the melt moves through the channels: the temperature rise is very approximately 1 °C per 1 MPa fall in pressure. Also, as the melt has become pressurized, there is an effect of pressure on viscosity. The effect is to increase viscosity, and this can be considered as equivalent to the increase in viscosity which would be caused by cooling, i.e. pressure can be regarded as an equivalent 'negative temperature'. The magnitude is again of the order of -1 °C per MPa pressure increase.

Some aspects of product quality

Thus, for changes in shear flow conditions, the effects of shear heating and pressure are opposite, and approximately cancel one another. As a rule, it may be said that greater error is introduced by allowing for only one or the other, than by ignoring both.

8.5.4 Orientation

One of the most important matters for injection moulders is orientation of the polymer as it enters the mould cavity and then freezes. In extruded products (Chapter 5), orientation is often desirable, leading to enhanced properties, but in injection moulding it is a nuisance. Usually, therefore, the aim is to minimize orientation, but this has to be balanced against the economic pressure to use rapid moulding cycles, which in turn implies rapid cooling of mouldings and consequent freezing-in of orientated distributions. If orientation, especially with crystalline polymers, is marked, there are frozen stresses which can cause mouldings to become distorted, either slowly, as stresses relieve spontaneously, or rapidly if elevated service temperatures are encountered.

What orientation patterns are we likely to find? A simple example would be an end-gated rectangular moulding. Figure 8.23 shows the pattern.

Fig. 8.23 Orientation pattern in a moulding.

1. When the melt enters the mould there is little orientation as the material contacts the mould wall; this leads to a low-orientation skin;
2. The bulk of the flow is laminar and highly orientated; inside the thin skin appears a highly orientated layer;
3. The centre may be less orientated because it remains hot, insulated by the outer layers, long enough to anneal.

In Fig. 8.24 we can see another orientation pattern, perhaps in a large flat moulding. At the constricted gate highly orientated patterns may result, under high stress conditions, and stress-induced crystallization may begin; this can initiate nuclei which will control the crystallization pattern in the moulding as it cools. Further in, the divergent flow can generate tensile hoop stresses which will cause warping or cracking in the moulding.

166 Injection moulding

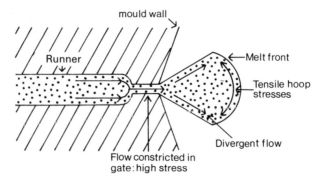

Fig. 8.24 Orientation effects at melt front.

8.5.5 Shrinkage

Another matter of great concern to the moulder is shrinkage. It is the difference in dimensions between the mould and the cooled moulding. The principal cause is the density change which occurs as the melt freezes. Crystalline polymers, e.g. acetal, nylon, HDPE, PET, polypropylene give the worst problems, with shrinkages of 1–4%. Amorphous polymers, e.g. polystyrene, acrylic, polycarbonate are easier, shrinking by only 0.3–0.7% (Table 8.2).

As we have seen for other problems, a combination of design factors and process control is used for the best results. Design factors include the

Table 8.2 Some approximate shrinkage values

Polymer	Percentage shrinkage
ABS	0.3–0.8
Acetal	0.0–2.2
Acrylic	0.2–0.8
Cellulose acetate	0.5
Nylon 6,6	1.5
Polycarbonate	0.6
Noryl	0.7
LDPE	2.0
HDPE	4.0
Polypropylene	1.5
Polystyrene	0.5
uPVC	0.3
Plasticized PVC	1.0–5.0

Some aspects of product quality

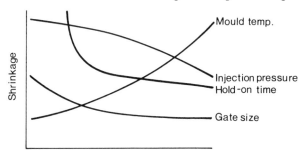

Fig. 8.25 Effects of processing conditions on shrinkage.

selection of a low shrinkage polymer if dimensional accuracy is important; filled grades, especially those containing glass are helpful. Sometimes this is not possible, e.g. for gears, which require the wear and frictional advantages of crystalline polymers. Precise dimensions are nevertheless required and it is necessary to 'aim off' slightly in the mould dimensions. To do this successfully the shrinkage has to be entirely predictable, and this requires symmetrical design in the product, and tool design which ensures even and symmetrical flow [5]. Usually, this implies maximizing the gate area and careful siting of gates.

Process control factors include minimizing mould temperature and the use of the optimum programme of injection and hold-on speeds and pressure. The maximum pressure is desirable to obtain rapid filling and maintain some hold-on pressure until the gate freezes, consistent with the avoidance of other faults. Figure 8.25 summarizes these factors.

8.5.6 *Injection profile*

For all moulding operations optimization of the injection cycle repays the effort of establishing it. At the least, optimization will ensure efficient material usage and low scrap rates; often it makes the difference between satisfactory and reject mouldings. A detailed case history of these effects can be found in ref. 3.

In outline, the injection process divides into two sections, viz. filling and packing, or hold-on. The filling section is speed dependent; ideally, filling should be rapid, so as to allow the check valve to operate quickly and positively. A modern programmable machine allows the speed to be varied, e.g.

rapid filling of sprue and runner system
slow down to avoid jetting through the gate

once the main cavity filling is under way, speed up again until full
further speed variations available to negotiate cores or other constrictions
in the mould.

This part of the programme relates *speed* to *distance*. At this point, the packing starts. The programme changes to *pressure* versus *time*. The correct pressure is used to maintain a tight mould fill but avoiding overpacking, which can lead to stressed and overweight mouldings. The packing pressure may be varied through the cycle to avoid flashing before the moulding surface freezes, then increased to tighten and eliminate voids, lowered to prevent stressing, until the gate freezes.

Such a programme may be set up for a new product, by experience and a few trial runs. For a modern microprocessor-controlled machine, the programme information can then be stored on disk or tape, ready for use on subsequent runs.

8.6 Sprueless moulding

One of the subsidiary but nevertheless important problems besetting the injection moulder is that of scrap arisings. The traditional moulds of the type we have studied above deliver a certain irreducible quantity of scrap in the form of sprues and runners. Added to this there will always be a certain number of defective mouldings which have to be scrapped. It is quite common to find moulds producing small objects in which the weight of sprue and runners exceeds that of the actual mouldings.

Now although thermoplastics are recoverable by regrinding and reprocessing this procedure is itself expensive in time and energy; the granulators consume considerable amounts of energy and are extremely noisy and therefore inconvenient to operate. Furthermore, the use of regrind material blended in with virgin stock means that a small amount of polymer will experience several processing cycles. In spite of the inclusion of stabilizers in the polymer, some degradation occurs, with consequent loss of properties.

For these reasons, modern mould design tends towards the elimination of large amounts of sprue material [7]. In the best of these the sprue is virtually eliminated altogether – sprueless moulding. The techniques include those listed below.

- Designing the mould so that the nozzle locates directly on the mould cavity. This is particularly suitable for large complex, single impression mouldings, e.g. vacuum cleaner cylinders;
- A similar type of technique extends the nozzle to locate directly without a sprue;

- Heated runners are now becoming commonplace in multi-impression tools. Electric heaters are built into the mould to keep the runners hot and the polymer in them molten. This melt is then used in the next shot;
- An alternative to heating the runners is to make them oversize. A layer of polymer on the wall then acts as insulation and the polymer in the centre remains molten. These are insulated runners.

8.7 Newer developments

Before leaving the subject of injection moulding, mention should be made of some derivative processes [8]. These are:

- structural foam moulding (SF)
- sandwich moulding (SM)
- reaction injection moulding (RIM) and RRIM, which is reinforced with short glass fibre
- the increasingly important use of thermosets in injection processes.

8.7.1 *Structural foam*

This development of the injection moulding process has developed for applications where stiffness is a prerequisite in the product. Examples include cases for electronic equipment – meters, controllers, computers, etc. – a trolley base for foodstuffs, a washing machine tank [9]. The way to increase stiffness in a component is to increase its thickness. There is a cube law relationship between thickness and stiffness. For a component of thickness s, material modulus E

$$\text{Stiffness} = Es^3$$

Increase thickness to $3s$

$$\text{Stiffness} = E \times 27s^3$$

As we have seen, there are difficulties in trying to mould thick sections, particularly sinking and voiding. The technique of structural foam moulding avoids these. In this development, the melt is 'expandable'; it contains a dissolved gas, or a chemical which decomposes at melt temperature to give a gas (usually azodicarbonamide which evolves nitrogen), and it expands to produce the foam when it leaves the pressurized injection unit and enters the mould. A 'short shot' of melt is injected into the cold mould, i.e. insufficient to fill the mould. This leaves space into which the foam can expand. We can see at once that there are a number of important differences from the conventional 'compact' injection moulding process, CIM.

1. As soon as injection of the short shot is complete, the valve to the injection unit is closed. Thus, the pressure of the expanding gas becomes the driving force to fill the mould. This is only about 3 MPa, compared with perhaps 140 MPa for CIM. Several advantages accrue from this;
2. The low pressure means that large projected area mouldings can be made using low closure forces;
3. Lightweight inexpensive moulds are possible, permitting shorter run lengths, although more durable steel ones may be preferred for long runs;
4. No hold-on pressure is required; the gas keeps the melt front moving;
5. The expansion and low pressure gives low-orientation products;
6. The expansion of the gas keeps the skin pressed gently against the mould surface. This prevents sinking and allows the ready moulding of thick sections for stiffness.

The foam is self-skinning; as it enters the mould and contacts the cold surface foam cells collapse to give the skin. Later material is insulated by the skin and retains its cellular structure. The surface has a characteristic rough, swirly finish, which is acceptable for many structural purposes, or which can be sanded and spray painted. However, there are applications for thick, stiff sections which also demand a perfect moulded surface. An example is a water cistern where the competitive products are in high glaze porcelain [9]. For these a different process for foam-cored moulding has been developed (sandwich moulding).

A more recent development of the structural foam principle has been pioneered by Peerless Foam Mouldings Ltd., under the name *Cinpres*. In this process, the gas is injected separately to form a large single bubble, i.e. a type of hollow moulding forms. This offers all the advantages of SF – thick sections, low closure forces, etc. It is claimed to allow lower overall density and hence greater weight savings than SF, with faster cycle times because the gas injection assists cooling.

8.7.2 *Sandwich moulding*

This technique injects skin and core melts separately. Obviously, this implies two separate injection units, and the plant is therefore complex and expensive. The skin polymer is normal and the core polymer contains the blowing agent. The two melts come together and co-inject concentrically at a specially designed nozzle. The procedure is

1. Start injection of skin;
2. Very soon afterwards admit the core melt, so that both melts are injecting together;
3. When the appropriate short shot has been injected, the core melt is shut off;

4. The skin material runs for a moment and then also shuts off;
5. The foam expands, the mould is filled and a perfect moulded surface forms on the skin.

As we saw for co-extrusion of packaging materials, a process closely resembling this one, there is little tendency for these viscous melts to mix. Instead the skin material behaves rather like a balloon being blown up by the expanding core. The sequence of injection – (skin); (skin + core); (skin) – ensures that there is a 'front' on the skin to be blown up. At the end, the cellular core is not exposed and the nozzle is primed for the next injection.

The sandwich moulding technique can be adapted to other uses as well as foam cores. The central core can be of a cheaper polymer or can be a scrap-containing layer. It can be used to confer special combinations of properties, e.g. for headlamps moulded from polybutylene terephthalate (PBT) in which the core is filled to give high temperature resistance and the skin is unfilled for a good surface finish. This process may be compared with injection moulded DMC headlamps (see Section 8.7.5). It is of course important to ensure that the two layers adhere well and have fairly similar shrinkage characteristics, otherwise the mouldings may fail or warp.

8.7.3 *RIM and RRIM*

The process of reaction injection moulding (RIM) and its derivative reinforced RIM (RRIM) differ from CIM by using reactive liquid resins instead of a polymer melt. There is therefore no extruder, but in its place is a system for storing and dispensing the reactive resins. Usually there are two components which undergo a spontaneous chemical reaction when mixed. They are metered and mixed immediately before injection into the mould (Fig. 8.26). Usually the chemical system is a polyurethane one, although there are some others under development. Polyurethanes and related systems like the polyureas are peculiarly suitable for the process, as well as being highly adaptable within themselves to permit variation for different purposes [8, 10].

The RIM process is carried out at relatively low temperatures. The storage and circulation systems usually require to be kept warm, say 60–90 °C. Circulation is continuous to maintain homogeneity in the components. When a shot is required a piston valve opens to admit metered amounts of the two components. There are two possibilities for mixing; in low pressure systems (used for e.g. shoe soles) a high-speed screw (15 000 r.p.m.) mixes the streams before injection; high-pressure systems have the whole circulation system under high pressure and mixing occurs by impingement of the two streams. In both systems some foam expansion is

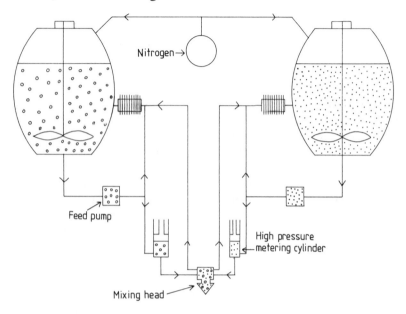

Fig. 8.26 RIM process.

used. Some products are flexible foams but others are expanded by only a small amount, say 10%, which allows the expansion to fill the mould in the way we have already seen for structural foam. This means low mould closure forces and the possibility of inexpensive moulds. An example may be found in the early days of development of car body parts in RRIM. Large, flat parts weighing 5 kg were being produced with a closure force of 50 tonnes. This is compared with the force required by other methods (Table 8.3). Overall, the low working temperature and absence of the expensive extruder, coupled with the lightweight press required mean that RIM plant is considerably lower in capital cost than that for CIM, SF or SM. It is the high pressure systems which have been developed in recent years for the production of automobile parts; initial interest was in shock-absorbing parts for bumpers, but more recently body panels and mouldings for internal instrument fascias have become important. There were two drawbacks in the early days; the mouldings were very different in thermal expansion properties from the steel parts to which they were attached, and they could not tolerate the temperatures for paint stoving without sagging. Reinforcement in the form of short glass fibres (< 2 mm) provided a useful improvement in both properties, and the RRIM process is gaining in importance in the field of body panels. Its superior impact resistance is

Table 8.3 Comparison of closure forces

Process	Closure force (tonnes)
RRIM	50
SMC compression moulding	800
CIM in polypropylene	2500
CIM in polycarbonate	3500

attractive for the wings of commercial vehicles, susceptible to minor impacts during loading operations.

8.7.4 Comparison of CIM, SF, SM and RIM

We have seen a number of advantages offered by the newer derivatives of the injection moulding principle. It is worth comparing the various methods, and asking why they do not supersede the traditional method of CIM altogether [8]. The main advantages offered are the ability of SF and RIM particularly to produce large projected area mouldings with low clamp forces and cheaper moulds. SM and SF come into their own for thick, stiff sections. However, CIM continues to be by far the most versatile method, easy to run and giving high output and trouble-free runs. The foam structures of SF and SM are insulating leading to slower cooling rates; RIM has the attraction of low capital cost, but it is limited largely to polyurethanes and it cannot reuse its scrap. As always in production processes, it is not possible to point to one which is superior in every respect; the production engineer must select the most economic and technically sound method for the requirements of each product.

8.7.5 Injection moulding of thermosets

Thermosetting polymers have traditionally been processed by compression moulding (Chapter 9), but injection moulding has been adapted more recently for these materials. They include the phenol–formaldehyde and urea–formaldehyde resins and also compounds based on unsaturated polyesters, especially dough moulding compound, DMC.

For thermosetting resins the injection moulding conditions are more or less reversed from those for thermoplastics. The screw is run at much lower temperatures, to avoid premature curing of the reactive resin and the mould is very hot, to cure the thermosetting resin quickly. Typical barrel temperatures are 75–80 °C for phenolics, about 100 °C for urea and up to 110 °C for melamine. The compression ratio of the screw is low, 1 or 1.1, to

174 Injection moulding

avoid local overheating from shear forces. DMC may well be run with a screw that is water cooled and no compression. It is not necessary to cool the mould before the moulding is ejected. An example is the moulding in DMC of vehicle headlamp reflectors [8, 11]. Increasing demands led to two dominant requirements in a reflector: (i) high temperature tolerance without distortion (up to 200 °C in some cases); (ii) the ability to mass produce reflectors with complex multiple curves in three dimensions. The second of these led to a choice of moulding rather than steel pressing. The temperature requirement led to a thermoset; note, however, the competitive method using sandwich moulding and thermoplastics (Section 8.7.2).

The material selected is DMC which is unsaturated polyester resin, syrupy in consistency, mixed with short glass fibre and dolomite filler. As the name implies it is doughy when mixed. It is force-fed into an injection moulding machine whose screw is cooled and injected into a mould at 180 °C.

In a general work on polymer processing, it is impossible to deal comprehensively with the vast topic of injection moulding. The account above is an attempt to summarize the most important facets of this complex and rapidly developing subject. The reader seeking more detailed and advanced information is recommended to one of the excellent works cited in the bibliography at the end of the chapter.

References

1. Crawford, R.J. (1981) *Plastics Engineering*. Pergamon, Oxford, Ch. 4.
2. ICI Ltd. (1976), Technical Service Note PP112, *Propathene for Integral Hinges* 3rd ed.
3. Morton-Jones, D.H. and Ellis, J.W. (1986) *Polymer Products*. Chapman and Hall, London, Ch. 2.
4. Morton-Jones, D.H. and Ellis, J.W. (1986) *Polymer Products*. Chapman and Hall, London, Chs 3, 4, 5, 6, 8, 9, 23, 24.
5. Morton-Jones, D.H. and Ellis, J.W. (1986) *Polymer Products*. Chapman and Hall, London, Ch. 4.
6. Austin, C. (1985) in *Developments in Injection Moulding – 3* (eds Whelan, A. and Goff, J.), Elsevier Applied Science, Barking, UK.
7. Rockenbaugh, R.E. (1985) in *Developments in Injection Moulding – 3* (eds Whelan, A. and Goff, J.), Elsevier Applied Science, Barking, UK.
8. Morton-Jones, D.H. (1985) in *Developments in Injection Moulding – 3* (eds Whelan, A. and Goff, J.), Elsevier Applied Science, Barking, UK.
9. Morton-Jones, D.H. and Ellis, J.W. (1986) *Polymer Products*. Chapman and Hall, London, Chs 7–10.

10. Morton-Jones, D.H. and Ellis, J.W. (1986) *Polymer Products*. Chapman and Hall, London, Ch. 12.
11. Morton-Jones, D.H. and Ellis, J.W. (1986) *Polymer Products*. Chapman and Hall, London, Ch. 18.
12. Bee, C.F., Technical Engineer, J. Fisher and Co. (Cleveleys) Ltd, private communication.

Further reading

Whelan, A. and Craft, J. (eds) (1978) *Developments in Injection Moulding – 1*. Elsevier Applied Science, Barking, UK.
Whelan, A. and Craft, J. (eds) (1981) *Developments in Injection Moulding – 2*. Elsevier Applied Science, Barking, UK.
Whelan, A. and Goff, J. (eds) (1985) *Developments in Injection Moulding – 3*. Elsevier Applied Science, Barking, UK.
Whelan, A. (1982) *Injection Moulding Materials*. Applied Science Publishers, Barking, UK.
Morton-Jones, D.H. and Ellis, J.W. (1986) *Polymer Products*. Chapman and Hall, London.

9

Compression and transfer moulding

9.1 Introduction

Compression moulding is the oldest mass production process for polymer products. We have mentioned it several times in previous chapters and it is appropriate to describe it at this point in the book to provide a comparison with injection moulding. It is almost exclusively used for thermosets, although these are also processed by the injection method, as we saw in Chapter 8.

The only important product to use a thermoplastic in compression moulding is the long playing gramophone record in black PVC copolymer. The main reason for this choice of process in this case gives a clue to one of the features of the process, namely, the low level of orientation in the mouldings. It is difficult to produce thin, flat discs like gramophone records by injection moulding without warping; the orientation leads to distortion to a dished shape or a saddle shape, both clearly quite unacceptable in this product. Compression moulding delivers the flat records which have been so successful for over 40 years, although now being overhauled in the market by the compact disc which is in fact injection moulded.

The compact disc is made from a specially developed grade of polycarbonate of low molecular weight to ensure good flow properties. In this respect it repeats the history of its predecessor the black disc for which a special PVC/PVAcetate copolymer was developed by the then British Geon company, to provide good moulding properties.

As a matter of interest, the compact disc also uses a special injection moulding machine with a screw capacity of only one shot, to regularize the thermal history of the moulding machinery. These factors illustrate the difficulty of mass-producing flat discs by injection moulding even to the smaller diameter of the compact disc.

9.2 Thermosetting compounds

9.2.1 *Resins*

The materials that form the main users of the compression moulding technique are the thermosetting resins and vulcanizable rubber. The latter

Thermosetting compounds

will be the subject of a later chapter; the former fall into the following main categories:

1. phenol–formaldehyde resins (the 'phenolics')
2. urea–formaldehyde resins
3. melamine–formaldehyde resins
4. epoxy resins
5. silicones
6. di-allyl phthalate and other alkyds
7. unsaturated polyesters (uPE).

No attempt will be made here to discuss the detailed chemistry of these materials; readers seeking such information are referred to the respective specialist texts [1, 2]. However, a general description of the salient materials and some of their common uses follows.

(a) *Phenolics*
The phenolics are among the earliest synthetic polymers to find widespread use. As the name implies, these materials are made by the chemical reaction of phenols and formaldehyde. They are available in many grades aimed at different markets. They constitute one of the oldest categories of synthetic polymer – bakelite, a name familiar for 60 years, was the first phenolic in large scale production. It was one of the earliest to make use of the inherently electrically insulating properties of polymers, in electric light fittings and plugs. It was also very widely used for the cases of 'wireless' sets of pre-war and immediately post-war vintage. It remains a widely used material today. Among the many uses of phenolics, we still find electrical fittings – plugs, sockets, light fittings, etc. The dark coloured examples of these items, usually a slightly mottled brown, are made from phenolic resins. These resins are naturally dark coloured and pale shades are not readily made. Phenolics give the cheapest electrical goods and they are satisfactory in electrical properties for general domestic applications. However, there is some tendency for 'tracking' to occur which renders them unsuitable for high tension use.

They also appear as dark, often black, handles on saucepans and kettles. A rather more exotic modern use was as the so-called 'ablative' shield on the nose cones of the Apollo space capsules. The shield insulated the capsule from the high temperatures generated on re-entry to the Earth's atmosphere, by virtue of its low thermal conductivity, and dissipated the heat by slowly burning away, without melting.

(b) *Urea–formaldehyde*
Urea–formaldehyde resins, made using urea in place of a phenol in chemical condensation with formaldehyde, are the materials from which the alternative white electrical fittings are made. Besides offering white or

coloured shades, they have better electrical properties than the phenolics, with less likelihood of tracking. The majority of domestic plugs and similar fittings are probably nowadays manufactured in these materials, although an increasing proportion of coloured 13-amp ring main plugs are made in injection moulded thermoplastics, such as nylon and PET.

(c) *Melamine resins*

Melamine–formaldehyde resins form the third important category of thermoset to be made from condensation of formaldehyde. The high chemical resistance of these materials has led to their widespread use in decorative laminates (*Formica* and similar products) and in robust tableware for use in camping and by children. They also possess outstanding electrical properties, but are the most expensive of the formaldehyde condensation resins.

(d) *Epoxies and UPE*

These materials when used for moulding are most often found in glass reinforced form. The epoxies offer better physical properties than the unsaturated polyesters, but are considerably more expensive. We shall meet these materials again in the chapter on fibre reinforcement. For conventional compression moulding, DMC is the most widely used uPE compound.

9.2.2 Compounds

The resins are blended with additives to make the final moulding compound, often supplied as a powder ready for charging the moulds. There are usually diluting or reinforcing fillers; these include fibrous fillers such as woodflour, cotton, asbestos, or glass, and granular fillers or flakes such as mica, talc, or slate flour.

The resin at this stage is, of course, 'incomplete', i.e. further polymerization and/or cross-linking is needed to bring it to its final chemical form, and this happens in the mould under heat and pressure. The moulding compound is thus a blend, made by distributive mixing (Chapter 3) of:

incompletely reacted resin
fillers, to reinforce or cheapen
catalyst, where needed, to promote the cross-linking reaction
accelerators, to modify reaction rate
lubricants, as processing aids and mould release agents
colourants and other special ingredients.

The 'pot-life', i.e. the length of time for which it can be stored, of such compounds varies:

SMC and DMC for days, up to 1–2 weeks
alkyds, polyesters without initiator, UF, for weeks to months
phenolics for up to 2 years.

Rubbers may be regarded as a 'sub-set' of the thermosets. They are compounded in the uncured state with various additives and are extensively moulded by compression moulding processes. They are the subject of a separate chapter, but share many features with the materials dealt with here.

Among the shared features of thermosets, including the rubbers, is the problem of volatiles which can present moulding difficulties. There are often volatile by-products of the chemical condensation reaction occurring in the mould. Frequently water is produced, and the fillers are often quite hygroscopic and contain appreciable amounts of water. The volatiles are the principal reason for adopting compression techniques in moulding these materials, to contain the volatile matter and prevent its 'gassing' to give porous mouldings.

9.3 Compression moulding process [3]

9.3.1 *Process description*

The principle of the compression moulding process can be outlined as follows.

1. The mould is held between the heated platens of a hydraulic press;
2. A prepared quantity of moulding compound is placed in the mould, usually by hand, and the mould placed in the press;
3. The press closes with sufficient pressure to prevent or minimize flash at the mould part line;
4. The compound softens and flows to shape; the chemical cure then occurs as the internal mould temperature becomes high enough;
5. If necessary, cooling takes place, although for the vast majority of thermosets this is not needed;
6. The press is opened and the moulding removed. Frequently, the mould is removed from the press and opened on the bench to extract the moulding. It is reloaded with a fresh charge before returning it to the press to commence another cycle.

In practice, the compound is often preheated to shorten the moulding cycle and to assist early flow in the mould. It may be taken to 60–100 °C by

infrared (IR) heaters
HF dielectric heating ovens (i.e. microwave ovens)

180 Compression and transfer moulding

a heated screw, which also compacts
convection heating in a hot air oven.

The moulding quantity is generally prepared by weighing or less frequently by volume, which is less accurate, or by dispensing from the heated screw in slugs of appropriate size. Alternatively, the powder is tabletted or pelletized in a tabletting press. A pressure of 90–125 MPa (6–8 tons per sq. in.) will tablet most solids. These preparations are the equivalent of the preparation of the shot of appropriate size by the injection unit in injection moulding. A slight excess of material is used, which flashes to a thin flash at the part line and rapidly cures to prevent further leakage.

The moulding cycle will often include a 'breathing' or 'bumping' sector; the pressure is momentarily relieved to release volatiles (trapped air and product gases) and bumped on again to 'puff' the gases out.

Temperatures are in the range 140–170 °C for the majority of thermoset resins, including rubbers. Thermoplastics, on the rare occasions of their compression moulding, need generally higher temperatures, e.g. 230 °C for polypropylene.

The moulding pressure varies according to the material. As with injection moulding, the requirement is to keep the press closed and this depends on the properties of the material being moulded, in particular its flow properties. Examples are shown in Table 9.1.

Table 9.1 Moulding pressures for thermosets

Material	Pressure	
	(mPa)	*(p.s.i.)*
DMC	6–10	900–1500
Granular uPE and soft flow phenolics	14–18	2000–4000
UF, MF, stiff phenolics	20–40	3000–6000
Stiffer materials	40–55	6000–8000

The pressures listed are the pressures applied to the moulding. They are found from the loading *force* and the moulding *area*. These should not be confused with the hydraulic *line pressures*, shown on some presses as gauge readings. The latter can be converted by observing ram size as well as hydraulic pressure, calculating the force and applying this to the mould area.

9.3.2 *Presses and moulds*

The original process was essentially manual, as described above, with loose moulds, and many such presses are still in operation. Control of cycles and temperatures was manual, with reliance on simple clocks and at best a

Compression moulding process

surface pyrometer for judging platen temperatures. Often, presses were steam heated and steam pressure as indicated by a gauge on the press was the only temperature control; in such cases the working specification would include a required steam pressure but these could be notoriously difficult to control with any degree of accuracy. In the same way, hydraulic pressure was observed manually. Such processes relied heavily on the skill and reliability of the operator; where wages depended on piecework output it was not unknown for quantity to be regarded as more important than quality.

The modern press has automated cycles, electrical proportionally controlled heaters and auto-controlled hydraulics. In many cases, mechanical mould filling and extraction is used. The advent of such modern presses was slow – indeed, as already stated, many old presses still operate. However, when used in a well designed shop, they allow important improvements in productivity; several presses can be supervised by one skilled operative, and the work becomes less manual and more technically challenging. The operative becomes more a technician than a manual worker.

The principle of the press is shown in Fig. 9.1, which depicts a simple manually operated press for clarity. Presses can be up-stroking or down-stroking, according to which platen is the moving one.

Three classifications of mould are identified, as shown in Figs 9.2, 9.3 and 9.4. The positive mould takes all the applied pressure; there is insufficient

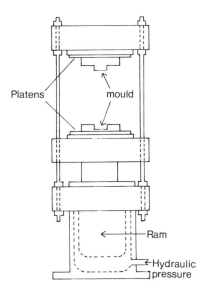

Fig. 9.1 Compression moulding press.

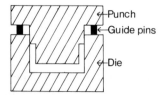

Fig. 9.2 Positive mould.

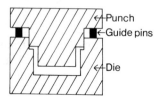

Fig. 9.3 Semi-positive mould.

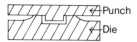

Fig. 9.4 Flash mould.

clearance to allow material to escape. Variation in charge weight in such a mould gives variation in thickness or density in the moulding. In general, positive moulds are not favoured. In the semi-positive mould the punch is arrested at a stop. Weighing of the powder is here only necessary to ensure a full shot on the one hand and avoid waste on the other. The simplification of the semi-positive design to the flash mould allows cheaper moulds and simplified operation. The flash forms the seal as it spews and rapidly cures, at the mould part line.

9.3.3 *Advantages of compression moulding*

The two features of compression moulding which distinguish it from injection moulding are:

1. The low scrap arisings (2–5%, but, it must be remembered, irrecoverable).
2. The low orientation in the mouldings; the product advantages that accrue from this are

    - fibrous fillers are well distributed and are not disturbed or orientated during processing

Transfer moulding 183

- the product has low residual stresses – hence the compression moulding of gramophone records, mentioned above
- mechanical and electrical properties are retained because there is little shearing flow to cause tracks
- mould maintenance is low; there is little erosion from the low shear forces, compared with injection moulding where tool erosion can be expensive
- capital and tooling costs are relatively low; plant and tooling are comparatively simple.

9.4 Transfer moulding

Transfer moulding is a development of compression moulding in which a reservoir of moulding compound is located in the mould and, upon closure, is transferred via runners to the cavities. Thus we see some relationship with injection moulding. The process is illustrated in Fig. 9.5.

Transfer moulding is used:

to give many small parts more easily;
to reduce the risk of damage or movement of thin or delicate mould parts or inserts;
because it is claimed to be faster due to better heat transfer through the runners.

In appropriate cases, transfer moulding offers these advantages; as we have seen previously, no single process is best for all products, and it is the job of

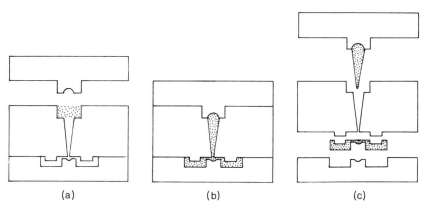

Fig. 9.5 Transfer moulding: (a) plug of moulding compound inserted; (b) press closes; compound transfers through runners to mould cavities; (c) three-plate assembly opens; mouldings recovered from lower daylight, sprue is withdrawn with top member.

Compression and transfer moulding

the process engineer and designer to find the most suitable process for the particular case. In reaching such a decision the disadvantages of transfer as opposed to straightforward compression moulding must be considered:

the flow usually gives unwelcome orientation in the product;
it also increases wear and maintenance costs;
tooling is rather more complex and hence more expensive;
the runners are scrap, with little chance of a hot runner system.

References

1. Brydson, J.A. (1984) *Plastics Materials*, 4th edn. Butterworths, London.
2. Saunders, K.J. (1983) *Organic Polymer Chemistry*. Chapman and Hall, London.
3. Monk, J.F. (ed.) (1981) *Thermosetting Plastics*. Godwin, London.

10
Polymers in the rubbery state

10.1 The rubbery state

As mentioned in the previous chapter, the rubbers constitute a 'subset' of the thermosetting polymers. An account of their specialized technology will occupy the next chapter: in this chapter a few generalizations concerning rubbery polymers are dealt with. In Section 1.4.2 we saw the relationship between structure and properties of polymers; the modulus–temperature diagram is reproduced in simplified form here, for ease of reference (Fig. 10.1). The rubbery state lies between the glassy and the molten states, and it is exhibited by all amorphous polymers to some extent. It is a plateau region of modulus, as shown in Fig. 10.1.

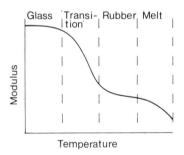

Fig. 10.1 Glassy, rubbery and molten states of an amorphous polymer.

When we speak of the rubbery state in processing it is important to realize that we do not mean a material which is 'snappy' and 'elastic' like a rubber band: such a material would, of course, be unprocessable, because it would recover elastically from any deforming process. This 'snappy' behaviour is caused by cross-linking of the polymer chains. In vulcanized rubber, the cross-links are chemical bonds, which are for practical purposes permanent. In *thermoplastic rubbers* the cross-links are physical and labile; a section dealing with the thermoplastic rubbers and their processing is included at the

186   Polymers in the rubbery state

end of Chapter 11. Most conventional (i.e. not classified as thermoplastic) rubbers also show this behaviour even before vulcanization and before their processing begins, but it is transitory and labile. The labile cross-links can be listed as follows:

- gel structures in natural rubber, before processing. These have their origin in the very high molecular weight (1 000 000) which leads to high entanglement;
- crystalline structures in polychloroprene (*Neoprene*)
- hydrogen bonding in polyurethane elastomers
- microcrystalline regions in plasticized PVC
- hard phase-separated segments in thermoplastic rubbers, e.g. styrene in SBS copolymers.

In the early stages of processing a rubber these temporary cross-links are ruptured by mechanical action and heat, a stage of processing known appropriately enough as 'warm-up' or 'break-down'. The result is a material of fairly similar modulus to the starting material, but with minimal elastic recovery, i.e. it is dough-like. Polymers which possess a long plateau in this rubbery region are suitable for the characteristic rubber processes. They include plasticized PVC as well as natural rubber and the custom-designed elastomers produced synthetically.

10.2   The calendering process

Calendering is a process uniquely applied to rubbery polymers, including plasticized PVC as well as compounded natural and synthetic rubbers. Because of its rather special place in relation to these two classes of materials it is described in detail here, rather than in the chapter on rubber technology.

Calendering essentially requires the polymer to be in the rubbery state. The process is the production of sheet of accurate gauge by passing the compound between rotating rolls. Often, more than one nip is required to give sheet of the required accuracy, and multi-roll machines are then used. Figure 10.2 shows some of the configurations adopted for these rolls.

Figure 10.2 shows the path of the rubbery polymer through the machine. The first sheet is formed at the first nip directly from the feed, which may be supplied from a two-roll mill or extruder. The second gap usually allows the sheet through without nipping. The sheet is then remade at the third gap, where a thin 'pencil' bank is kept rolling. Under good conditions, the gauge of this sheet can be controlled to $\pm$ 0.02 mm. The inverted L design would be preferred for thick sheet because it gives a long dwell for full heating. At the other extreme the inclined Z offers a short heat history for thin or heat-sensitive sheet. Fairly clearly, plastics and rubber calenders are similar to

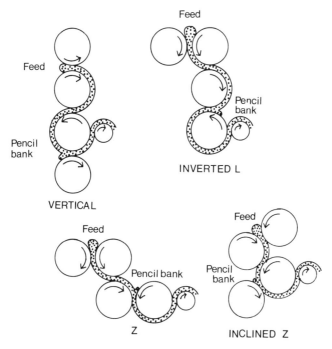

Fig. 10.2 Common roll configurations for four-roll calenders.

two-roll mills in principle. Their rolls are held a fixed (but adjustable) distance apart, i.e. they do not 'float'. Contrary to widespread belief, these machines do not derive from the machines called 'calenders' in the textile industry. The textile machines are more correctly termed 'rotary presses': they are constant pressure machines, whose rolls will float. Plastics and rubber calenders are constant gauge machines in which the forces to maintain the gauge may vary depending on the material being processed.

The forces generated in this way between the rolls may be considerable. Thus, for a sheet of 100 μm gauge, 30–40 ton, and for 50 μm, 70 ton separating forces may be generated. This is sufficient to bend the rolls and cause the gap between them to become convex, which gives a sheet of uneven gauge, edge to edge. Such a sheet is unsatisfactory for most purposes because it distorts when reeled up; when unrolled again it will reveal a baggy centre. Roll bending is compensated to restore a parallel gap in processing (Fig. 10.3) by:

(a) grinding a compensating, opposite contour on the roll
(b) pre-loading the ends of the rolls to give compensatory bending
(c) slightly crossing the roll axes to give thicker edges.

188   Polymers in the rubbery state

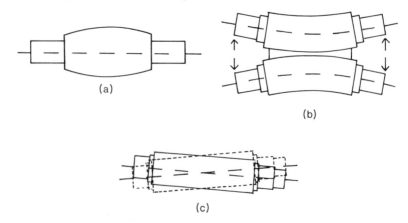

Fig. 10.3 Methods for compensating for roll bending: (a) roll contouring; (b) preloading; (c) crossed axes.

Of these methods, contour grinding is the simplest and the only method if the machine lacks the facilities for roll bending or crossed axes, but is the least versatile. A given contour will usually only be satisfactory for a narrow choice of materials and thicknesses, and regrinding is a laborious and often expensive procedure. It is important not to over-compensate, to give a concave sheet; this will also distort on reeling and the sheet will have wavy edges on unreeling. The provision of roll bending or cross axis facilities increases capital cost of an already costly machine: however, most modern machines are likely to have one or even both of these provisions.

Calenders are used for the production of sheeting. For rubber goods, this is afterwards assembled into various configurations for many products, e.g. tyres, belting. PVC sheeting more often finds direct application in its own right, e.g. shower curtaining, rainwear and other apparel, tarpaulins, and in thicker gauges as floorcovering for contract installations.

Rubber calenders run with roll temperatures of about 110°C which maintains good rubbery-state processing without risk of premature vulcanization. PVC calenders are run at temperatures approaching 200°C, when the plasticized PVC behaves as a rubber – it is really the first thermoplastic rubber. Figure 10.4 shows a typical layout for a PVC calendering plant. The raw materials enter at the start and a few minutes later finished sheeting is being reeled up at the end of the line.

Calendered sheet is usually highly orientated, and exhibits anisotropic behaviour in its physical properties. The process is, of course, highly linear. The polymer is aligned along the sheet, in the machine direction. There are important tensile components as the sheet leaves the nip, and

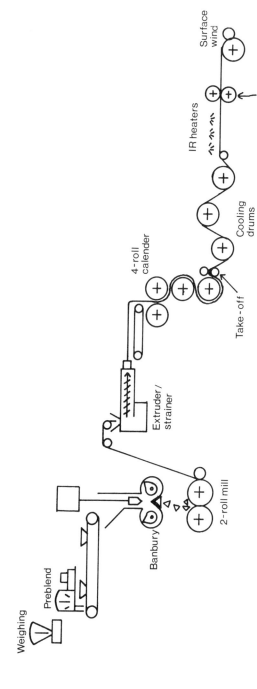

Fig. 10.4 PVC calendering plant.

predominantly shear forces within the nip. The best calendered sheet surface is usually found to be a matt finish, even with polished rolls. It is considered that this matt finish is similar to the 'sharkskin' effect in extrusion, and is caused by very small scale tensile rupture as the sheet leaves the nip.

11
Rubber technology

11.1 Types of rubber

Rubbers can be classified in several different ways. We have already seen that there are thermosetting and thermoplastic rubbers. The thermoplastic rubbers form a special classification, at least from the processing standpoint, because they are processed by conventional thermoplastic machinery: they are discussed at the end of this chapter. The thermosetting rubbers are the main concern here. Their processing, test procedures and products constitute the oldest industry to use polymeric materials, originally entirely natural rubber but now extended to include many types of synthetic elastomer. (The words 'rubber' and 'elastomer' may be regarded as more or less interchangeable; 'elastomer' is a technical word which describes the material to some extent; 'rubber' is a more colloquial word which derives from one of the first uses of natural rubber, as an eraser.)

Thus, rubbers are classified also as 'natural' or 'synthetic' and the synthetics are themselves sub-classified by chemical type or by application, e.g. resilient or oil-resistant or flame retardant, etc. A few of the most prominent chemical types are listed below, with notes on their most important features.

(a) *Natural rubber, NR (approved ASTM abbreviation)*
Chemically, natural rubber is polyisoprene

$$\left[-CH_2-\underset{\underset{CH_3}{|}}{C}=CH-CH_2- \right]_n.$$

It has excellent resilience, and low hysteresis properties. The hysteresis is a measure of the energy absorbed when the rubber is deformed. In a tensile test it may be seen as the loop enclosing an extension and the return curve as the extension is reversed (Fig. 11.1). The absorbed energy is equivalent to the reciprocal of the resilience, i.e. low hysteresis≡high resilience, and high hysteresis≡low resilience. Low hysteresis rubbers are used in applications where low energy absorption is important, e.g. tyre walls,

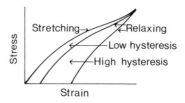

Fig. 11.1 Hysteresis loop in a tensile experiment.

where low energy absorption prevents heat build up as the walls flex. Conversely, high hysteresis rubbers are used for tyre treads where the low resilience and energy absorbing properties reduce bouncing and thus help grip on the road. Natural rubber is a general purpose rubber still very widely used; it constitutes approximately 30% of total rubber usage.

Processes now exist for the production of 'synthetic natural rubber', by the controlled polymerization of isoprene monomer. The first of these was Goodyear's *Natsyn*.

(b) *Styrene butadiene rubber, SBR*
SBR is the highest volume rubber of all, with about 60% of the total. This figure taken with the 30% for NR means that all the others together comprise 10%. SBR is a random copolymer of styrene and butadiene; when a block copolymer is made the thermoplastic elastomer known as SBS is formed. The proportions of styrene and butadiene vary for different purposes; the large volume, general purpose rubber is 23% styrene, 77% butadiene. This is a relatively high hysteresis rubber which finds application in tyre treads.

Higher proportions of styrene are used in special 'reinforcing' grades which are used in blends to improve toughness and abrasion resistance; these may have 50% or higher proportions of styrene, which makes them much harder and more resinous.

(c) *Butadiene rubber, BR*

$$\left[CH_2-CH=CH-CH_2 \right]_n.$$

The chemical structure of this elastomer can be seen to be the simplest of the range of diene rubbers, NR has a methyl (CH_3) group in place of a hydrogen atom. All the diene rubbers are polymerized from diene monomers, which leaves an unsaturated double bond still present in the polymer molecule.

BR is the lowest hysteresis rubber of all. It exhibits very high resilience (rebound). It is made by solution polymerization methods which lead to a narrow molecular weight distribution. Because of this it is rather difficult to process on its own and is usually used in blends.

(d) *Butyl rubber, IIR (isobutene–isoprene rubber)*
Butyl rubber is made not from a diene monomer but from isobutene which has only single unsaturation; the polymer is thus saturated. To provide some sites of unsaturation for vulcanization 0.5–3.0% isoprene is copolymerized with it. Butyl rubber shows the opposite set of properties to butadiene rubber. It exhibits high hysteresis and very low rebound, although the resilience becomes comparable with that of NR at 100 °C, at which temperature the mobility of its substituent methyl groups increases

$$\left[-CH_2-\underset{\underset{CH_3}{|}}{\overset{\overset{CH_3}{|}}{C}}- \right]_n .$$

This polymer should not be confused with polybutylene, made from normal butylene, which is a polyolefin related to polyethylene and polypropylene

$$\left[-CH_2-\underset{}{\overset{\overset{CH_3}{|}}{\underset{|}{CH}}\underset{}{_2}}-CH- \right]_n .$$

(e) *Chloroprene rubber, CR*
This is another diene rubber. A chlorine atom substitutes for the hydrogen in butadiene

$$\left[-CH_2-\overset{\overset{Cl}{|}}{C}=CH-CH_2- \right]_n .$$

The best known rubbers of this class are the Neoprenes manufactured by DuPont. The chlorine renders them self-extinguishing, hence their use in conveyor belting for coal mines [1]. It has moderate resilience and some oil resistance.

(f) *Nitrile rubber, NBR*
Here the hydrogen of butadiene is replaced by a nitrile group, CN, to give acrylonitrile, which is then copolymerized with butadiene. Nitrile rubber is thus an acrylonitrile–butadiene copolymer. It is the most oil resistant of the bulk commodity rubbers, finding application in seals and automobile parts which come into contact with mineral oil. There are different grades of

NBR, depending on nitrile content. High nitrile contents (35–40%) have the best oil resistance but are more expensive, have higher T_g and hence poorer low temperature properties and are stiffer. Medium (25%) and low (18%) nitrile content grades thus find application where oil resistance is less important

$$\left[-CH_2-\underset{\underset{CN}{|}}{C}=CH-CH_2- \right]_n .$$

(g) Ethylene–propylene rubbers, EPM rubbers, ASTM class 'M'

EP rubbers are copolymers of ethylene and propylene. They contain 60–80% ethylene, the principal function of the propylene being to prevent crystallization of the ethylene. The polymerization method is chosen to give a random copolymer. A fully alternating copolymer is obtained by hydrogenating (i.e. saturating the double bonds with hydrogen) natural rubber. EPM is the only polymer suitable for insulation of 60 kV power cables; only oiled paper offers better insulating properties.

A third monomer is frequently copolymerized in the EP system to confer unsaturation on the molecule and hence make it capable of vulcanization. This third monomer has to be a diene, as in the case of butyl rubber mentioned above. The resultant rubber is designated EPDM (ethylene–propylene–diene monomer). There are variations in the diene selected, but ethylidene norbornene is frequently used

Notice the resultant unsaturation is not incorporated in the main polymer chain. This gives good resistance to weathering and particularly attack by ozone, which is a long-standing problem with the diene rubbers.

11.2 Production of rubber

11.2.1 Synthetic elastomers

The synthetic rubbers are, of course, products of the petrochemical industry. They are produced through chemical routes for monomers which are shared by other polymers, some of which are outlined in Chapter 1. Unlike polymers destined for other processing routes, such as extrusion, blow moulding, injection moulding, compression moulding, synthetic elastomers are usually supplied not as granules but in large 100 kg bales to match the traditional bales of natural rubber for which the industry is equipped.

11.2.2 Polymer latices

Natural rubber is tapped from the tree *Hevea brasiliensis* as natural latex. A latex is a colloidal dispersion of polymer particles in water, i.e. a solid-in-water colloid or sol in which the solid or disperse phase particles are macromolecules.

Other polymers are supplied as latices for use in the so-called emulsion paints and industrial surface coatings and adhesives. The emulsion polymerization route, in which polymer is made from an emulsion (oil-in-water colloid) of monomer, produces latex polymer which has to be dried to give solid polymer. When required, the latex can be retained. Polymers manufactured by other routes can be converted to latices, often by precipitating from solution under prescribed conditions, in the presence of surfactants and stabilizers to prevent coagulation of the colloid.

Natural rubber latex is used directly for some processes. Dipping is the most used. The product is made by dipping a shaped former into the latex and precipitating a layer of solid polymer on to it; an example is the production of rubber gloves.

Natural latex is used, in concentrated form, as the binding backing in tufted carpets. The tufts of yarn are not anchored to the backing fabric, usually woven jute hessian or polypropylene, as they are in a woven carpet, and the function of the latex is to provide the anchorage. After tufting, latex is roller-coated on to the back of the carpet and the coated material passed through a drying oven to remove the water from the latex, which results in the tufts becoming firmly anchored in a tough layer of rubber. The same concentrated latex is used as a domestic adhesive, under the names '*Copydex*' or '*Revertex*'.

Natural latex contains approximately 30% solids, and about 10% of it is used as latex. The concentrated latex referred to above is prepared from this variously by centrifuging, evaporating and creaming to 60% solids. Ammonia (0.2%) is added to stabilize the concentrated product.

11.2.3 Solid natural rubber

The bulk of natural latex is converted to *ribbed smoked sheet*, *RSS*, the form in which natural rubber is normally supplied. To make this, the following stages are followed.

- The latex is diluted to 15%;
- It is coagulated with formic acid and stored for 1–18 h to mature the coagulum;
- The coagulum is pressed through rollers to remove the bulk of the water, to a 5 mm sheet. The final rollers have grooves which give a characteristic criss-cross pattern to the sheet;

- The sheets are dried by the smoke from burning rubberwood – hence the name ribbed smoked sheet. The smoke contains natural fungicides which prevent mould growth.

A superior grade called *pale crepe* is made by a modified route. The latex is diluted to 20% and then fractionally coagulated to remove the fraction containing the yellow β-carotene pigment, or alternatively it is bleached. The formic acid coagulation and milling follow, but warm air is used for drying rather than smoke.

There are other variants and about 25 grades exist. In recent years the RSS grading system has been replaced with one designating the grade of *Standard Malaysian Rubber, SMR*.

11.3 Vulcanizing

11.3.1 *Introduction*

The first uses of natural rubber, crudely recovered from the latex or applied directly, as waterproofing, revealed some limitations. It became stiff and brittle at low temperatures and soft and sticky when hot. The invention of *vulcanization* overcame these problems quite dramatically, allowing the development of many new uses. Not the least of these was the application to vehicle tyres, especially through the parallel invention by Dunlop of the pneumatic tyre.

We know that vulcanizing processes cross-link the polymer chains, to reduce chain slippage and hence stabilize the morphological structure. Several chemical routes are used including peroxide cross-linking and the use of metal oxides for chloroprene rubbers, but the mainstay of the industry is still sulphur vulcanization. A modern vulcanizing formulation contains not only sulphur but also several other chemicals which control the speed and smoothness of the reaction. The best way to examine these is to follow the historical development of the process.

11.3.2 *Development of vulcanization*

The reaction of rubber with sulphur was discovered independently by Goodyear (1839) in Woburn, Massachusetts and Hancock (1843) in London. This first vulcanizing mix was very simple:

NR   100 pbw (parts by weight)
sulphur   8 pbw
cure: 5 h at 142 °C.

By today's standards, the properties of this compound would be rather deficient. It represented a considerable advance, however, at the time. The next improvement was to introduce zinc oxide as an activator. This chemical forms an intermediate with the sulphur and this speeds up the reaction

NR   100 pbw
sulphur   8 pbw
ZnO   5 pbw
cure: 3 h at 142 °C

Further improvement came early in the twentieth century when the first organic accelerator was used. Aniline was the first material to be effective but was toxic. Its reaction product with carbon disulphide, thiocarbanilide, was the first to be used extensively:

NR   100 pbw
sulphur   8 pbw
ZnO   5 pbw
thiocarbanilide   2 pbw
cure: 1.5 h at 142 °C

The first true accelerator came in 1921, when mercaptobenzthiazole, MBT, was applied. This accelerator is still very widely used

[structure of mercaptobenzthiazole: benzene ring fused to a thiazole ring with C–SH]

Its function is to form a complex with the sulphur and the zinc oxide, and the effects are:

1. A reduction in the amount of sulphur needed, which means that the sulphur is being used more effectively. In the early formations the sulphur formed bridges consisting of several sulphur atoms between polymer chains. The more efficient and smoother reaction using MBT leads to shorter sulphur links which in turn gives better properties and better ageing characteristics;
2. A shorter cure time is achieved;
3. The 'bin life' of the compound before curing is extended.

Research at this time also revealed the importance of natural fatty acids, often already present in the rubber, to the efficiency of accelerators in maintaining an acid environment. Stearic acid is now always added as standard and is especially needed in synthetics. Our formulation has now become:

NR   100 pbw
sulphur   3 pbw
ZnO   5 pbw
stearic acid   1 pbw
MBT   1 pbw
cure: 20 min at 142 °C.

Rubber technology

The remaining problem was oxidation of the mix during processing. These mixes were made by open mixing on a two-roll mill, which constantly exposed fresh surface to attack by atmospheric oxygen, at an elevated temperature, say 60–80 °C. Antioxidants were introduced with great effect. They are usually sterically hindered phenols and amines (i.e. phenols and amines with bulky substituents on the aromatic ring which electronically and physically hinder their chemical properties). These antioxidants interrupt the free radical chain reaction by which oxidative degradation of polymers, including rubbers, proceeds. Our final formulation is thus:

NR	100 pbw
sulphur	3 pbw
ZnO	5 pbw
stearic acid	1 pbw
MBT	1 pbw
antioxidant	1 pbw.

This formulation remains a very serviceable one today. The main advances have been a huge range of new accelerators (Table 11.1), used in combinations which act synergistically, and the emergence of the so-called *efficient vulcanization*, *EV* systems (see below).

The most important of the new accelerators are the ultra-accelerators, which are used in conjunction with an accelerator like MBT to give more rapid and complete cure. Another important class of accelerators is comprised of those with a delayed action. These do not become effective until a threshold temperature is reached, after which they are fast and efficient. They permit safe processing before vulcanization, followed by a rapid cure.

One of the problems which is always a concern of the compounder is reversion. The vulcanization reaches an optimum stage, as monitored by an appropriate test, and then starts to reverse, i.e. the monitored property starts deteriorating. The complex chemistry of sulphur vulcanization is still not fully understood, but reversion must result from the fact that, like many chemical processes, it relies on pushing an equilibrium in the desired direction: it is easy for it to pull back again!

The same principles and ingredients, with slightly modified proportions, are applied in NBR, SBR, IIR, EPDM and BR. Cross-linking may also be achieved in other ways.

1. Peroxides are used for EPM and the speciality silicone and fluorocarbon rubbers;
2. Magnesium and zinc oxides are used for chloroprenes; the metal reacts with the chlorine to leave an oxygen bridge. Chloroprenes are also vulcanized by sulphur and accelerators, depending on the grade of rubber in use.

Vulcanizing 199

Table 11.1 Some examples of organic accelerators

Name and chemical formula		Initials	Features
Diphenylguanidine	(C₆H₅-NH)₂C=NH	DPG	Medium-fast
Di-o-tolylguanidine	(CH₃-C₆H₄-NH)₂C=NH	DOTG	Medium, disperses better than DPG
Mercaptobenzthiazole	benzthiazole-C-SH	MBT	Widely used, fast, scorchy
Dibenzthiazyl disulphide	(benzthiazolyl)-S-S-(benzthiazolyl)	MBTS	Less scorchy than MBT, preferred in SBR
N-Cyclohexylbenzthiazyl sulphenamide	benzthiazolyl-S-NH-C₆H₁₁	CBS	Delayed action, fast
Zinc diethyldithio carbamate	$\left(\begin{array}{c} C_2H_5 \\ C_2H_5 \end{array} N-\overset{S}{\underset{\parallel}{C}}-S^- \right)_2 Zn^{++}$	ZDC	Ultra-accelerator
Thiuram sulphides $\begin{array}{c} R \\ R \end{array} N-\overset{S}{\underset{\parallel}{C}}-S-S-\overset{S}{\underset{\parallel}{C}}-N \begin{array}{c} R \\ R \end{array}$	R = methyl, tetramethyl thiuram disulphide	TMT or TMTD	Ultra-accelerators
	R = ethyl, tetraethyl thiuram disulphide	TET or TETD	Sulphur donors in sulphurless cures
$\begin{array}{c} R \\ R \end{array} N-\overset{}{\underset{\parallel S}{C}}-S-\overset{}{\underset{\parallel S}{C}}-N \begin{array}{c} R \\ R \end{array}$	Monosulphides	TMTM TETM	

11.3.3 Some practical vulcanizing mixes

Table 11.2 shows a range of representative vulcanizing formulations [2]. The progress of a vulcanization can be followed experimentally by changes

Rubber technology

Table 11.2 Vulcanizing formulations for different rubbers

Ingredient	NR (pphr)†			SBR (pphr)			NBR (pphr)			IIR (pphr)		EPDM (pphr)	
	*												
ZnO	6	5	5	3	4	4	2	5	5	5	5	5	2.5
Stearic acid	0.5	1	0.5	3	2	2	3	1	1	2	2	–	0.5
TMTD	–	–	0.04	–	–	–	–	–	–	1.5	1.5	–	–
MBT	0.5	–	–	–	–	–	–	–	–	0.5	0.5	0.5	0.25
ZDC	–	–	–	–	–	–	–	–	–	–	0.75	–	–
MBTS	–	–	0.25	–	1.2	–	–	1.5	1.5	–	–	–	–
DPG	–	–	–	–	–	0.8	–	–	–	–	–	–	–
CBS	–	0.4	–	2	–	1	2	1.5	–	–	–	–	–
TMTM	–	–	–	–	0.15	–	–	–	0.4	–	–	1.5	0.75
S	3.5	3	2.5	2	2	1.75	2	1.5	1.5	2	2	1.5	0.75
Cure time (min)	40	30	20	30	25	25	30	25	20	40	30	25	55
Cure temperature (°C)	141	141	141	153	153	153	153	153	153	153	165	165	165

* This is the Standard Compound of the American Chemical Society, used as a reference formulation.
† Formulations are usually expressed as 'parts per hundred of rubber', pphr, rather than as percentages.

in some selected physical property. If tensile testing is used, the pattern of variation shown by tensile strength, TS, modulus, E, and elongation at break, EB, for different types of rubber will be different. Figure 11.2 shows the difference for NR and SBR. Natural rubber (Fig. 11.2(a)) reaches maxima for tensile strength and modulus; these maxima are at different cure times. Following the maxima comes the phenomenon of reversion, and the

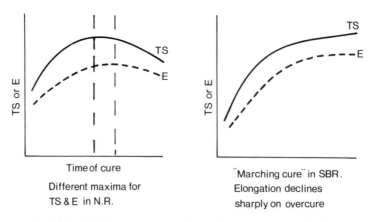

Fig. 11.2 Different cure curves for different rubbers.

Fillers 201

properties begin to decline again. Clearly, the compounder must decide which property is of prime importance and determine an optimum cure cycle which must then be strictly observed.

SBR, on the other hand, shows an initial rise in strength and stiffness to an optimum level. After this there is slower continued cure, the so-called 'marching cure', characteristic of this material. Over-cure is betrayed in SBR by a sharp decline in the elongation at break (not shown).

An important development in sulphur vulcanizing technique has been the emergence of *efficient vulcanizing* (*EV* systems). These formulations use low sulphur content, or even no elementary sulphur at all, relying on S-donors such as the thiurams and thiocarbamates which will supply the sulphur through the intermediate complex. These systems have a greater proportion of mono- and di-sulphide links, compared with the polysulphides of some of the more old-fashioned compounds. Examples of comparative formulations are shown in Table 11.3.

Table 11.3 Examples of comparative formulations

	S-cure (pphr)	EV-cure (pphr)
RSS	100	100
ZnO	5	5
Stearic acid	2	2
HAF black	35	35
MBTS	1.25	–
TMT	0.15	–
Sulphur	2.5	–
CBS	–	1.5
ZDC	–	1.5
SBR	100	100
ZnO	5	5
Stearic acid	2	2
HAF black	60	60
MBTS	1.25	–
TMT	0.15	–
Sulphur	2.5	–
ZDC	–	1
TMT/TET blend	–	1

11.4 Fillers

Fillers in rubber are finely divided solids added to the rubber during mixing. They are divided into reinforcing and non-reinforcing types, depending on whether they enhance the properties of the rubber or act simply as extenders. Far and away the most important is carbon black, the dominant reinforcing filler.

11.4.1 Carbon black

(a) *Basic characteristics*

This material has been known for thousands of years as a black pigment for inks and paints. About 1915, S.C. Mote discovered the superior abrasion resistance of tyres incorporating C-black.

All processes for the production of C-black use the same principle. Elementary carbon is deposited from the vapour phase at very high temperatures as the result of thermal decomposition of hydrocarbons. It is not made by carbonizing organic substances. The process produces a smoke from which sooty particles deposit. The characteristics are

1. Aggregated particles of colloidal dimensions and high surface area;
2. The atoms are arranged in a 'quasi-graphitic' formation, i.e. aromatic ring systems but with less order, size and perfection than classical graphitic structures;
3. Primary aggregates are fused from much smaller particles, and vary from clustered, grape-like aggregations to more bulky, branched, filamentous forms. Two important properties are determined to characterize different classifications. These are:

    (a) *Structure*: for a constant surface area, the number of particles or nodules within each aggregate, and
    (b) *Surface area*.

The structure is determined by the *dibutyl phthalate absorption test (DBPA)*. The liquid DBP is dropped on to a weighed sample of black and mixed, in a mixing instrument designed for the purpose, until no more can be absorbed, as shown by a sudden increase in torque required by the mixer. This finds the voids volume, which is related to structure. The surface area is determined by an iodine adsorption test in which iodine is adsorbed on the surface quantitatively. The range of values found for these properties in rubber grade C-blacks is:

Structure: 60–180 cm^3 $(100 g)^{-1}$
Iodine adsorption: 20–270 $mg\,g^{-1}$ ($\simeq$ 20–270 $m^2\,g^{-1}$).

The exception is the grade called medium thermal black, which has DBPA of 30 cm^3 $(100 g)^{-2}$ and I_2 adsorption of 7 $mg\,g^{-1}$.

(b) *Manufacture*

The two most important processes for manufacturing carbon blacks are the furnace process and the thermal process. Also used for minor quantities are the channel or impingement or lampblack process, and the acetylene process.

The *Furnace* process uses as its feedstock residuals from the aromatic stream in petroleum refining. This is injected into a high velocity stream of combustion gases from the complete burning of a fuel with excess air. A portion of the feedstock burns to maintain the flame temperature, but most cracks to carbon and hydrogen. There is 50–70% yield of carbon, dependent on the surface area. The structure is adjusted by trace amounts of alkali metals. The plant is a cylindrical refractory lined furnace, 2–7 m long and 15–90 cm in diameter. The temperature is 1200–1700 °C. The residence time is of the order of seconds or milliseconds. Downstream, there is a water quench to 1000 °C, then a series of heat exchange systems. The product is collected on bag filters as a fluffy powder at 240 °C. This filtration temperature necessitates heat-resistant filter cloths, which are made variously from glass, *Nomex* or *Teflon* woven cloth. *Nomex* and *Teflon* cloths have a raised and needlefelted surface to give the correct permeability to filter the colloidal-sized black particles from the hot gas stream.

The *Thermal* process cracks natural gas thermally, in the absence of air, at 1300 °C. It gives the largest particle size blacks, with low structure and surface area [3]. It gives 40–50% available carbon yield.

The *Acetylene* process is a variant of the thermal process which uses acetylene feedstock; once the reaction has started it maintains its temperature through the exothermic reaction.

The *Channel* process was the major one until superseded by the furnace process. Thousands of small natural gas flames impinge on iron channels to deposit the carbon, which is scraped off. The yield is low, about 5% carbon being recovered. Channel blacks have a surface area of about 100 mg g^{-1}, with normal to low structure. Their combustion method of production leads to an acidic, oxidized surface which gives good processing in rubber, with slower vulcanization, hence scorch resistance and good reinforcement. They are, however, expensive, because of their inefficient production, and the production method is polluting.

(c) *Classification*

The old system classified carbon blacks according to various performance criteria. It designated them by a lettering code, and this system is still in use, even though it is inconsistent and confusing. Some of these classifications refer to abrasion resistance produced in the rubber compound.

   HAF – high abrasion furnace
   ISAF – intermediate super abrasion furnace
   SAF – super abrasion furnace.

Some refer to reinforcement

   SRF – semi-reinforcing furnace.

204   Rubber technology

Some to vulcanizate properties:

HMF – high modulus furnace.

Some to processing:

FEF – fast extrusion furnace.

Some to particle size:

FF – fine furnace.

There are others, and in addition subdivisions exist such as GPF (general purpose furnace).

Table 11.4 Carbon black classifications

Type	Symbol	ASTM	Mean size (μm)	Surface area ($mg\,g^{-1}$)	pH
REINFORCING					
Super abrasion furnace	SAF	N110	14–20	120–140	9–10
Intermediate super abrasion furnace	ISAF	N219 N220 N231 N242 S315 N326	18–24	110–120	8.5–9
High abrasion furnace	HAF	N327 N330	24–28	75–95	8–9
Hard processing channel	HPC		22–25	100–110	3.7–4
Medium processing channel	MPC		25–29	90–105	3.8–4
Easy processing channel	EPC		29–33	80–90	3.8–5
MEDIUM REINFORCING					
Fast extrusion furnace	FEF	N550	30–50	45–70	9–10
Fine furnace	FF	N440	40–45	55–70	9–10
General purpose furnace	GPF	N660	50	45	9–10
High modulus furnace	HMF	N601	45–65	30–60	9.5–10
Semi-reinforcing furnace	SRF	N761 N762 N770 N774	60–85	25–45	9.5–10
Lampblack	LB		100–150	13–25	4
CONDUCTIVE					
Superconductive furnace	SCF	N294	16–20	120	9.5
Conductive furnace	CF	N293 N296	24	110	9
Conductive channel	CC		17–23	100–150	3.5–4
Acetylene	–		40–45	40–70	7–9
OTHERS					
Fine thermal	FT	N880	120–200	15–35	8.5–9
Medium thermal	MT	N990	250–500	5–10	7–9

Fillers 205

A more systematic nomenclature is gaining ground; this is the ASTM 'N' and 'S' categories, of which 100 is the finest and 900 the coarsest, with sub grades within the hundreds. 'N' signifies normal cure rate and 'S' slow. Eighty per cent of black usage is found in three major grades:

HAF (N300)
GPF (N600)
FEF (N500)

Table 11.4 shows the main classifications.

(d) *Effect on compounds*
Carbon blacks greatly modify the physical performance of rubber compounds. The effects they exert are dominated by the two properties already discussed, surface area or 'particle size', and structure or 'aggregated size'. High surface area (small particles) imparts:

high levels of reinforcement, leading to high tensile strength
resistance to abrasion and tearing.

High structure (large aggregates) imparts:

improved extrusion behaviour
higher stock viscosity
improved 'green strength', i.e. strength before curing
high modulus, i.e. stiffer products

It is not possible in this general work on polymer processing to include detail of these effects. The reader requiring fuller information should turn to ref. 3.

(e) *Mechanism of reinforcement*
During mixing, the black is rapidly 'incorporated' (distributive mixing). Its voids are filled with rubber. Dispersive mixing then follows; high shear forces break down the agglomerates until final dispersion is reached. Often, a two-stage process is used to mix. In the first stage a concentrate of rubber and black, called a *masterbatch*, is made (see below); the DBPA test gives the maximum black loading possible in the masterbatch.

If the compound is examined by solvent extraction, the so-called *bound rubber* content can be determined. This is the insoluble portion and forms an interconnecting system of chains and particles, as a fragile black gel, e.g. a compound of rubber (100 pbw) and black (50 pbw) will give perhaps 35% bound rubber. Clearly, there is powerful interaction between rubber and black particles. The particles may act as 'stress homogenizers', allowing slippage and a redistribution of stress among the polymer chains. Other mechanisms exist as well: mechanical rupture of polymer chains, during processing, gives free chain ends which can react with free radicals on the black surface, with a result equivalent to a giant cross-link, at the morphological rather than the molecular scale. The reinforcement effects are most marked in intrinsically weak, non-crystallizing rubbers like SBR, NBR, EPDM. Crystallizing rubbers like NR are already strong; reinforcement may still be important, but its criteria are less clear. Examples are:

1. A pure gum SBR vulcanizate may have a typical tensile strength of 2.2 MPa. Addition of 50 pphr of reinforcing black would raise the TS to 25 MPa.
2. Pure gum NR would typically show a stress of 45 MPa at 700% extension; with 50 pphr SAF black only 550% would be possible, and the stress would be 35 MPa. However the abrasion resistance of the NR compound would be improved.

Thus, tensile strength is not a universal criterion. Stress at 300%, a frequently quoted property for rubbers, would be no criterion either; many inert fillers would raise this without improving any failure property.

The best single criterion is the *energy to rupture*, illustrated in Fig. 11.3. In this Figure we can see the increase in rupture energy when black is used, reaching a maximum at the optimum loading and then falling again as excess functions as an inert filler. When barytes is used there is no reinforcement and the rupture energy falls throughout.

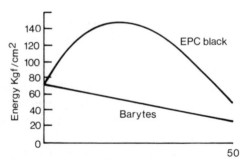

Fig. 11.3 Energy of rupture as a criterion for reinforcement.

11.4.2 *Mineral fillers*

A number of powder minerals are used as fillers, and a few are listed below.

(a) *China clays*

These are chemically hydrated aluminium silicates. There is a range of grades characterized by different particle sizes. For rubber use, they are graded:

Soft, >2 μm, semi-reinforcing

Hard, <2 μm, reinforcing
Calcined (heated to remove bound water), reinforcing.

The reinforcing action of clays is much less pronounced than that of C-black. It is useful in compounds where a black colour is unacceptable.

(b) *Calcium carbonate*

The two main types are ground limestone and precipitated calcium carbonate. The first is made by grinding mineral limestone and the second is obtained by chemical precipitation from solution. Both are extenders only and offer little reinforcing effect.

(c) *Silica*

1. Ground mineral sand, below 200 mesh, is a non-reinforcing filler.
2. Hydrated precipitated silica, containing 10–14% water, and having a particle size range of 10–40 nm, is a reinforcing filler.
3. Fume or pyrogenic silica, having more than 2% water and/or very fine particle size, is a highly reinforcing filler. It will give light-coloured products with reinforced properties, but it is expensive and brings processing difficulties; its compounds with rubbers work-heat rapidly and tend therefore to be scorchy.

Various other minerals, e.g. barytes (barium sulphate), are used in smaller amounts as inert fillers.

11.5 Processing methods

11.5.1 *Overall processing route*

The processing routes used in the production of rubber items uses individual processes already described in earlier chapters. The overall route may be summarized in the diagram in Fig. 11.4.

11.5.2 *Mixing*

Rubbers are almost invariably mixed in internal mixers such as Banbury or Intermix machines. The action and principles of these have already been discussed in Chapter 3.

Rubbers have high viscosity, which becomes even higher when reinforcing fillers are added, with the result that 150 °C can easily be reached. Such a temperature would be disastrous if vulcanizing agents were to be added. The result would be the rubber processor's nightmare, a scorched, or prematurely vulcanized batch. A scorched batch is a total loss, and there is the need to clean up afterwards. As we have seen, the solution is to employ

208 Rubber technology

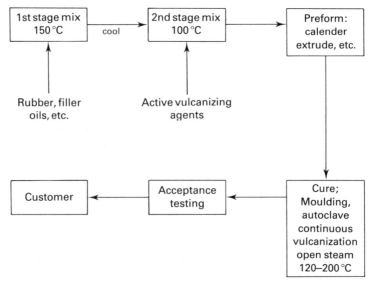

Fig. 11.4 Overall processing route for rubbers.

two-stage mixing. The rubber and black, together with other fillers and extending and softening oils which the compound may contain, are mixed in the first stage. This mix is discharged from the mixer and cooled before proceeding to stage 2. The second stage is more easily kept cool and the vulcanizing agents are added here. The cooler mixer also allows completion of disperse mixing of fillers, because of the high viscosity at lower temperatures. The mixing consumes considerable quantities of energy, as we have seen in Chapter 3. It is common for a first stage Banbury mixer to be driven by a 1000 h.p. motor.

The high shear characteristic of rubber mixing is essential to the breakdown of black agglomerates, and to the dispersive mixing of many other ingredients, which, as we have seen, often need to act together. A typical mixing schedule is given in Table 11.5 for a natural rubber mix in a 100 kg mixer. Dumping is usually on to a two-roll mill, which easily accepts the large masses of compound discharged from the Banbury. The mill cools it and converts it to an easily handled sheet: alternatively a strip can be taken from the mill to feed an extruder, which itself might feed a calender, or could pelletize the compound. The pellets would be dusted with whiting or calcium stearate to prevent sticking.

One of the problems encountered in rubber mixing is the accurate handling of the very large numbers of separate ingredients. Most rubber formulations require vulcanizing, stabilizing, plasticizing and filling additives, and different formulations require different additives and

Table 11.5 A typical mixing schedule for a natural rubber mix

Time (min)	Temperature (°C)	Stage
Stage 1		
0	70	Add rubber and small powders, i.e. not fillers or vulcanizing agents. Ram down
2	90	Rubber now broken down. Add half filler, ram down
4	150	Add remaining filler and oils
6	150	Ram up, dump on mill, sheet off and cool.
Stage 2		
0	60	Add master batch
2	90	Add sulphur and accelerators
3	100	Dump

different quantities. Also, a particular problem surrounds the use of carbon black, because its intensely black nature makes it an especially dirty material to handle. This leads to difficulties with contamination of non-black mixes, and it is also unpleasant and dirty for the operatives. The most modern mixing plants use mechanical material handling plant to store and dispense ingredients. Fillers, including blacks, are stored in bulk in silos. Ingredients are weighed and dispatched to the mixer from a central control console, which indicates the stages to the operative. In the most up-to-date installations, the details of all recipes are held in a computer data bank, the computer also controlling the mixing and discharge cycle. A recently installed computer controlled installation at the Dunlop Precision Rubber plant can handle 150 separate ingredients, including 10 types of carbon black. Three hundred recipes can be stored with up to 26 ingredients each. The mixing programme can give 42 instructions. End points can be preset for time or power consumed. The silo storage of black, coupled with efficient extraction ensures a good working environment and avoids contamination.

11.5.3 *Preforming*

The most commonly used preforming processes are calendering and extrusion, both of which have been discussed in earlier chapters. They produce accurate sheet or profile. Calendered sheet is mainly intended for building into products immediately prior to vulcanizing, e.g. tyres, conveyor belting.

Extruders for rubber are similar in principle to those for thermoplastics. The temperatures are lower (ca. 100 °C), and usually somewhat less critical. L/D ratios are lower (10–15) as are compression ratios (1.4:1). These differences reflect the different requirements of rubber extrusion and especially the requirement to avoid scorch: there is nothing worse in polymer processing than a scorch in an extruder! Before feeding to a calender or extruder, the rubber stock needs to be warmed up, usually on a mill. If freshly mixed, it is already in the correct condition, but if it is drawn from a mixed rubber store, it will have regained its 'nerve' and hence the need to restore it to the viscous, less elastic state. In extreme cases a *cracker mill* may be used. This machine is a two-roll mill whose rolls have diagonal grooves: it is powerful and very rapidly breaks down a nervy rubber stock.

When correctly warmed up mixed rubber forms good, even gauge sheet or accurate profile. There is still an elastic component, of course, as evidenced by extruder die swell and its calendering equivalent 'spring-back' – the sheet gauge is somewhat greater than the roll separation – but the deformation is predominantly viscous.

11.5.4 *Thick calendered sheet*

The production of thick sheet has always presented problems. The attempt to produce it directly from a calender results in air inclusions in the body of the sheet as well as surface blisters, because of insufficient shear in the centre of thick sheet. The traditional solution has been to ply up several thicknesses of thin sheet, but this procedure is laborious and troublesome, and also expensive. An alternative has been to extrude a tube, slit it and lay flat the sheet: this is limited to low outputs and widths (500 kg/h, 1000 mm).

A combination machine pioneered a few years ago by the Berstorff company gave a new way to produce thick sheet. This is the roller die extruder. The extruder delivers a 30–60 mm thick slab directly into a calender nip, across its full width. The die is a coathanger design (Chapter 4)

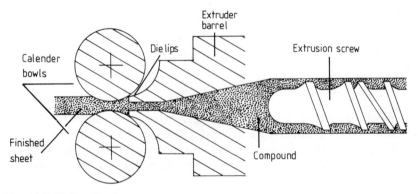

Fig. 11.5 Roller die plant.

giving consistent slab material. The calender then reduces the gauge to that required, without a rolling bank to mill in air. The calender is thus an extension of the die (hence the name 'roller die'). A pressure sensor in the die lip controls the calender speed to keep the stock pressure in the die region constant, thus matching the outputs of the two units. The machine is illustrated in Fig. 11.5.

11.5.5 Vulcanizing

In most cases, vulcanizing requires heat and pressure. The temperature is usually in the region of 150 °C to make the vulcanizing reaction proceed at the correct speed and to a satisfactory completion. Pressure is needed to contain volatiles evolved during the reaction; without it the product is porous and gassy. The only exception is thin surface coatings, where the volatiles can escape.

(a) Compression moulding

This is the most widely used technique. The procedure is basically as described in Chapter 9 on thermosets – rubber is of course a thermoset. However, the rubber is not as a rule supplied to the press in weighed aliquots of powder. Rather, it is as calendered sheet or an extruded preform. For complex items like tyres, the product is often built from many different components – a tyre contains over 50 individual components. All these are assembled before the whole is pressed in the mould, which is itself in a press. The cure time for large and complex items may be 20–30 min, for simple small ones, 5 min. For items like tyres, each is individually moulded. Other products, like conveyor belting, can be made in continuous lengths by pressing a section at a time in a press with open ends [4] (Fig. 11.6). It is usually best in rubber moulding if the preform or 'blank' is a rough fit for the mould. This avoids stresses and orientations which readily deform a low modulus material like a rubber.

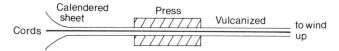

Fig. 11.6 Intermittent press for a conveyor belt.

(b) Rotary presses

Figure 11.7 shows a rotary press of the Rotocure type. This has a continuous flexible steel belt passing round a large heated drum. Pressure is maintained through the belt–drum contact zone. The drum is 1.5–2 m in diameter and the speed of the machine is such as to give a dwell in the pressure zone sufficient to cure, about 15 min being typical. In the 'Auma' machine made

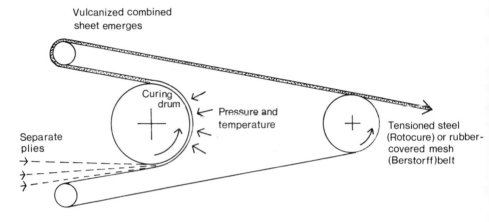

Fig. 11.7 Rotary vulcanizing press.

by Berstorff the belt is of woven wire mesh covered with a rubber compound. In Rotocure and Berstorff machines, the return run of the belt carries the cured, plied product (e.g. rubber/textile conveyor belting) to a wind up station, cooling as it goes.

(c) *Open steam cure*
The conventional way to use steam to cure is to place items to be cured in an autoclave, on suitable racks or carriers, and then to admit steam to the required pressure, which gives the appropriate temperature and pressure for curing. The heat transfer from steam is efficient because of its very high heat capacity.

An alternative way of using steam is found in the manufacture of fire hose. The extruded, unvulcanized hose is pressurized with steam. The pressure starts low, to avoid blowing through the soft unvulcanized rubber; as curing gets under way the inside toughens up and pressure can be gradually increased to a maximum of perhaps 80 p.s.i. The hose thus cures from the inside out.

(d) *Continuous vulcanization (CV)*
A few products lend themselves to continuous vulcanization in tandem with extrusion. Examples are covered cable and high pressure hydraulic hose. The extruded product is admitted through a seal into high pressure steam – about 200 p.s.i. The dimensions of the product and the good heat transfer allow the extrusion and curing to proceed in tandem. The problem with CV for most products is the marked disparity between possible calendering or extrusion speeds and curing rates.

11.6 Testing

Three important areas of testing can be identified for rubber goods, and they will be discussed in turn. They are:

- compound 'plasticity', which is essentially viscosity testing
- curing characteristics
- properties of the finished vulcanizate.

11.6.1 *Plasticity and curing characteristics*

These two properties are discussed together because they are frequently determined at the same time, using the Mooney or Monsanto plastimeters. For fundamental research-based work, the capillary extrusion rheometer is used, but for process and quality control functions the *Mooney plastimeter* is the most widely used instrument. In this instrument, a disc of compound is rotationally sheared. The instrument's disc rotates within a shallow cylindrical cavity, and the space between the disc and the cavity wall is filled with the test rubber. The rubber is squeezed in with considerable pressure. The disc is the rotor, the cavity walls are the stator. Both are serrated to afford a grip (Fig. 11.8). Two disc sizes exist; the larger is 1.5 in. in diameter

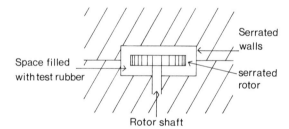

Fig. 11.8 Disc and cavity of Mooney plastimeter.

and is used for most work; the smaller, 1.2 in., is used only for very high viscosities. The cavity is contained within thermostatically controlled, heated platens. The disc rotates at 2 ± 0.02 r.p.m. The usual temperature is $100 \pm 0.5\,°C$, but other temperatures may be selected for special purposes.

The procedure is:

- 0 min   load sample and close platens
- 1 min   start rotor (the first minute is for warm up)
- 4 min   read plasticity.

The torque is balanced against a U-shaped spring in the machine. The reading obtained in this way is recorded as

$$\text{ML}\,(100\,°C)\,1 + 4 = (\text{reading})$$

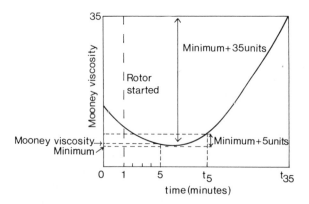

Fig. 11.9 Scorch and cure characteristics by Mooney plastimeter.

which means Mooney, large disc, 1 min warm up, 4 min rotation. The viscosity/time plot from the Mooney plastimeter yields further information after the 1 + 4 reading has been taken. Figure 11.9 shows a typical curve. The plot shows that the curve reaches a minimum some time after the 1 + 4 min position (i.e. 5 min total). It then starts to rise rapidly as cure gets under way.

The *Mooney Scorch Time*, T_5 is the total time, from platen closure, to when the plasticity reading is the *minimum + 5 units*. T_{35} is the time to cure, (minimum + 35) units, and $T_{35} - T_5$ ($= T_{30}$) is the *cure index*. The most frequently quoted readings are ML 1 + 4 and scorch time (T_5).

Another widely used instrument for following vulcanization rates is the *Monsanto curemeter*. A typical cure curve is shown in Fig. 11.10. This machine also measures plasticity but uses an oscillating disc instead of the constantly rotating one of the Mooney. The minimum is recorded as the plasticity of the uncured rubber. As Fig. 11.10 shows, a 'fractional modulus'

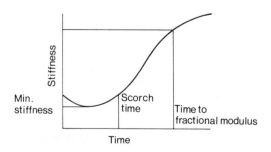

Fig. 11.10 Curve from the Monsanto curemeter.

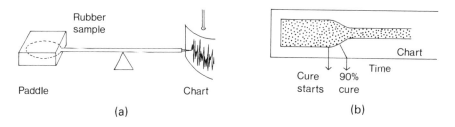

Fig. 11.11 Shawbury curometer principle and chart.

is noted for the full cure time; this is a selected fraction of the maximum possible, often 90%, and represents a technically achievable cure, as well as being easier to determine. The scorch time is determined at a fixed increase above the minimum, indicating incipient cure.

A third instrument which records the progress of a cure is the *Shawbury curometer*. A paddle oscillates in a small block of (initially) uncured rubber with a sinusoidal motion, and this motion is transmitted to a recording chart. The rubber is heated and as cure begins the oscillations of the paddle are progressively diminished. A characteristically-shaped trace appears on the chart (Fig. 11.11) on which the cure time is readily found.

Finally, mention should be made of *variable torque rheometers*. The best known is the Brabender Plasti-corder, but there are similar machines available from RAPRA/Hampden. They are essentially miniature processing machines; the heads can be changed to simulate various processes. The most frequently used are miniature Banbury-type mixing chambers, but extruders and others can be supplied. In these machines, the reaction torque is converted through an array of levers, or in more modern examples, electronically, to a read-out display: essentially, the power to drive the device against its contents is recorded, and the progress of a mix is followed by its changing plasticity.

11.6.2 Product testing

Rubber products vary so much that no general limits on properties can be stated. The common characteristics of rubber products are:

rubbery elasticity, in tension and compression
low modulus, in shear and tension
flexibility.

There are hundreds of special tests designed to evaluate particular products, but there are a few tests that are very widely used and understood by rubber technologists universally. These include the following.

(a) *Tensile testing*

A specimen, often dumb bell shaped, is held in the jaws of a tensile testing machine. It is extended at a known steady rate and the force developing in it is registered by the machine. The test is usually continued until the specimen breaks. If the cross-sectional dimensions of the specimen are known, the force can be converted to stress and a stress/elongation curve obtained. From this various parameters can be derived. The most commonly quoted are:

> The stress at 100% extension;
> The stress at 300% extension. (These two values are often referred to as 'moduli', but, of course, they are not moduli);
> The ultimate tensile strength, expressed as a stress reading.
> The elongation at break.
> The initial modulus is also often found as a tangent from the origin of the stress/strain graph;
> Other moduli are found as secant moduli, by drawing the secant from the origin to a designated elongation on the curve and finding its slope;
> The area under the curve, either to break or to some specified stress or strain limit, gives a measure of toughness, and is useful in comparative experiments;
> A variant is to extend the specimen to a predetermined point and then to reverse the drive to obtain a recovery curve. There is usually a loop between stressing and recovery curves which represents the hysteresis energy.

A modern machine will be interfaced with a computer which considerably enhances its power as an analytical tool, enabling previously long or tedious determinations to be performed rapidly and with improved accuracy. An outstanding example is the machine developed by Nene Instruments, which is driven entirely from the computer, with no controls on the machine itself, and with automatic calculation of a large array of properties.

(b) *Set testing*

When a stretched or compressed specimen is released, its original dimensions may not be fully recovered. A tension or compression set test determines this as a percentage of the original deformation. A very commonly used test is compression set in which original thickness of a specimen is measured. The specimen is compressed between steel plates by 25% of this original thickness, using spacers. It is maintained in the compressed state under prescribed conditions, often 22 h at 70 °C. It is then released and its final thickness measured after 30 min recovery. The test as described here is notorious for its irreproducibility. This is because a single point is selected in a long-term test. The test is really a creep experiment and there is much to be said for recognizing this and recording the set values at a series of times and plotting the results to give a set versus time curve. If log axes are used a nearly straight line emerges, where slope is characteristic. Alternatively, the projected time to, say, 90% set can be used to compare materials.

(c) *Hardness*
The hardness of a rubber is measured as depth of penetration of a spring loaded indenter. It is related to Young's modulus. There are two scales in common use. Shore A for soft and Shore D for hard rubbers. Both record hardness on a scale of 0–100; when readings of A exceed about 95, D will begin to yield more useful comparisons. There is another scale, roughly equivalent to Shore A, called IRHD (international rubber hardness degrees).

(d) *Tear strength*
A number of standard tear strength specimen types are specified for different purposes. Testing for tear strength is normally done using a tensile testing machine. The curves generated during a test are often complex, as the specimen tears in a ragged manner, and their interpretation depends on the detailed specification being followed.

(e) *Others*
The above sections summarize a few of the most commonly required general tests. There are many others aimed at more specialized requirements. A few of these are listed below:

   resistance to chemicals, by immersion in them
   resistance to oils and hydraulic fluids
   flexing to failure
   abrasion resistance
   retention of properties at low or elevated temperatures
   low temperature flexibility, i.e. is the rubber below its T_g and therefore brittle.

11.7 Thermoplastic elastomers

This class of elastomers, also called thermoplastic rubbers (TPE or TPR), emerged in 1965 with the development of styrene–butadiene–styrene (SBS) block copolymer. As we have already seen, this copolymer differs greatly from the random copolymer, which is SBR. The styrene and butadiene components are incompatible so that when they are present in blocks they form separate phases, joined at the junctions between blocks. The result is a butadiene rubber, or 'soft segment' effectively cross-linked by 'hard block' styrene.

These cross-links are as effective in their influence on properties as chemical ones such as those made by sulphur, except that they are thermally labile. There are two results of this: the TPRs can be processed as thermoplastics, and, by the same token and in common with all thermoplastics, there is a ceiling temperature for service which is lower than that for chemically cross-linked rubbers. Nevertheless, SBS has made its mark and is widely used, e.g. for boot and shoe soling; most rubber bands are now made from SBS.

Other chemical types have followed. Prominent among them has been a copolymer of the polyester polybutylene terephthalate (PBT) as the hard segment and the polyether poly tetramethylene ether glycol (PTMEG) as the soft segment. This copolymer type, copolyether–ester, is marketed as *Hytrel* by DuPont and *Arnitel* by Akzo. It is made in a number of copolymer compositions and in melt blends to offer a range of hardnesses for different applications. These materials are also widely used, e.g. for conveyor belting, snowmobile tracks.

A number of other TPEs may be mentioned. Among the most versatile are the thermoplastic polyurethanes. These are finished polyurethane polymers whose chemical make-up gives thermoplastic properties. (Other classes of polyurethanes finish their polymerization actually in the mould, e.g. in the RIM process and in the formation of polyurethane foams.)

The thermoplastic polyurethanes are made from polyether or polyester soft segments coupled to short hard segments by urethane linkages. Thus they also are block copolymers with phase separated hard segments cross-linking rubbery soft segments.

Perhaps the first thermoplastic rubber of all is plasticized PVC, although it was not recognized as such in its early days. Nevertheless, it has rubbery properties and is processed by thermoplastic routes, e.g. for shoe soling.

The TPEs are normally processed by extrusion or injection moulding. This processing route was a hurdle in the early days of these materials. Rubber processors, in whose market sector most applications lay, had no equipment suitable for them, and the thermoplastics processors who were fully able to process the TPEs had no outlets for the products. For a time, the potential of these exciting new materials fell between these two rather separate parts of the polymer processing industry. The equipment used is very similar in type and operation to that for the more conventional thermoplastics.

References

1. Morton-Jones, D.H. and Ellis, J.W. (1986) *Polymer Products*. Chapman and Hall, London, Ch. 21.

2. Morrell, S.H. (1982) in *Rubber Technology and Manufacture*, 2nd edn (eds Blow, C.M. and Hepburn, C.) Butterworths, London, pp. 171–201.
3. Horn, J.B. (1982) in *Rubber Technology and Manufacture*, 2nd edn (eds Blow, C.M. and Hepburn, C.) Butterworths, London, pp. 202–63.
4. Morton-Jones, D.H. and Ellis, J.W. (1986) *Polymer Products*. Chapman and Hall, London.

12
Fibre reinforced plastics

12.1 Introduction

This is another sector of polymer processing which exists virtually as a separate industry, with its own specialist practices. One or two of its characteristic materials have already been mentioned in previous chapters, especially resin–glass systems and these are discussed further below.

We are concerned in this chapter with the materials made from a matrix resin and long reinforcing fibres. Although short fibre reinforcement is widely used today in thermoplastics and thermosets like DMC, and in RRIM, its function is somewhat different from that seen in long fibre reinforcement, often being essentially an increase in modulus and fracture toughness. These properties are shared by long fibre reinforcement, but there are additional property and processing characteristics which set this industry and its products apart.

12.2 Materials

12.2.1 *Fibres*

The most widely used reinforcing fibre by far is glass. The term 'GRP' (glass reinforced plastics) is often applied to this section of the industry and its products.

The glass most used is E-glass which is an acid, borosilicate glass like Pyrex. C-glass is more chemically resistant and S-glass has higher strength and modulus but is more expensive. E-glass is the 'workhorse' of the whole fibre reinforced plastics industry (FRP).

Other important fibres are carbon (CFRP), *Kevlar* and some specialist inorganic fibres. Carbon fibre is made by carbonizing an organic fibrous polymer, usually polyacrylonitrile, under specialized conditions. *Kevlar* is an 'aramid' fibre developed by DuPont, but also produced now by other organizations under other trade names. Aramids are aromatic polyamides, and are thus related to the nylons. They use aromatic starting materials, whereas the ordinary nylons are aliphatic. As its simplest, *Kevlar* may be regarded as the polyamide derived from paraphenylene diamine and terephthalic acid (i.e. the *para* acid). It is thus linear and inherently strong. *Kevlar* is a yellow fibre, somewhat unstable to light. Its strength is suggested when one tries to cut a woven cloth made from it with a pair of scissors.

Another aramid fibre may be mentioned here in passing: it is not used in reinforcement, but is of interest because of its relationship to *Kevlar*. This is *Nomex*, which is made from metaphenylene diamine and isophthalic acid (the *meta* acid). *Nomex* does not have the strength of *Kevlar* and it is supplied as staple fibre. It is extremely heat resistant and is widely used in filter bags for filtering hot gas streams, at above 200 °C (see Chapter 11, manufacture of carbon black), and for racing drivers' overalls, because of its heat resistant properties and the fact that it does not melt: this property has also led to its use for firemen's tunics in some brigades, allowing more comfort than the conventional wool melton. It is alleged that *Nomex* clothing will withstand a 'flashover' fire.

The important criterion in selecting a fibre for reinforcement use is that its modulus must exceed that of the matrix it is to reinforce, i.e. the fibre must be stiffer than the resin. The Young's modulus (E) of the polyester and epoxy resins mostly used in this work are in the range $6-8 \times 10^3$ MPa. For textile fibres such as PET (terylene) and nylon, $E \simeq 9 \times 10^3$ MPa, about the same as the resin, and they do not reinforce. They are, however, effective in reinforcing rubbers because the stiffness of rubbers is a good deal lower than that of the hard, brittle polyesters and epoxies. Glass and aramid fibres have values of E in the range $1-9 \times 10^5$ MPa. For carbon fibre, $E \simeq 10^6$ MPa, and these fibres confer powerful reinforcement.

Fibres are used in a number of different formats. They can be in continuous lengths in *rovings*, rather like an untwisted yarn. The rovings can be woven into *glass cloth*, in which the usual variations in weave construction can be used (plain, twill, satin, etc.), or they can be simply laid unwoven. Glass cloth gives reinforcement in linear and cross directions, linearly laid rovings only in the linear direction. Alternatively, continuous fibres may be laid down in a swirl pattern, or chopped fibres laid in a random manner, and the mat bonded with a resinous binder, to give continuous fibre mat and chopped strand mat. In chopped strand mat, the fibres are chopped to about 6 cm length. These mats give isotropic properties.

12.2.2 *Resins*

The most widely used resins are the *unsaturated polyesters*, upe. The epoxy resins [1] are also used for more demanding applications, but are more expensive.

The upe resins are made chemically from three types of starting material:

- an unsaturated acid, i.e. one with a double bond in its molecule; a commonly used one is maleic acid;

222   Fibre reinforced plastics

- a saturated acid, often phthalic acid;
- a glycol, i.e. a molecule which has a hydroxyl group on each end, often propylene glycol.

Both acids react to give a viscous, syrupy, mixed ester which contains sites of unsaturation. Styrene monomer is added: at first it acts as a thinner, giving improved flow and spreading properties. When a free radical initiator is added the styrene and the unsaturated sites on the upe polymerize. The styrene effectively cross-links the upe chains and the hard, infusible finished resin results (Fig. 12.1). There is considerable latitude for controlling the rate and temperature for the curing process, by the selection of the initiator. Those used for repair kits and hand lay-up work are usually active at room temperature, and the reaction begins as soon as the resin and 'hardener' (initiator) are mixed. There is enough time to perform the application before the reaction renders the mixture too stiff to handle. For hot curing, as in SMC and DMC (see below) a different initiator is selected which is

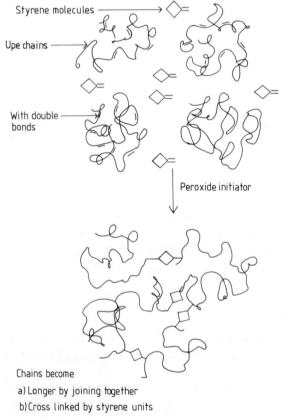

Fig. 12.1 Formation of polyester resin.

nearly inactive at room temperature but becomes very active at curing temperature. For example, tertiary butyl perbenzoate, TBP, is used in SMC intended for hot press moulding.

The activity of an initiator like this can be measured by its *half-life*. It acts by splitting its molecule at the O—O peroxide linkage to form two free radicals, and it is these that initiate the polymerizing chain reaction. The half-life measures the time for half the molecules to split. TBP has a half-life of 1 min at 166 °C and 10 min at 141 °C. Details of SMC preparation appear in Section 12.5.

A recent development is the use of carbon fibre reinforced *PEEK* in aerospace components. PEEK is *poly ether ether ketone*. It is a high temperature thermoplastic which is supplied in thin sheets already impregnated into linearly laid carbon fibre, *prepregs*. The moulder plies the prepregs to the required thickness and compression moulds to shape, usually a panel component. The temperature required is about 400 °C, and the resultant composite has outstanding strength and temperature tolerance. The moulding process, however, is not without its difficulties, and the product is expensive.

More conventional thermoplastics and glass fibre can be combined similarly to give a type of prepreg, by rolling together the melted polymer and long-staple glass, after passing through an oven. The resultant prepreg is cut into preform shapes for compression moulding. Typical products are load-bearing floor pans for commercial vehicles. These materials are marketed as *Azmet*, using PET as the polymer and *Azdel* using polypropylene, by GE Plastics.

12.3 Mechanical strength of fibre reinforced composites

There are two aspects to the mechanical strength properties of these materials, which to some extent make conflicting demands on formulation. These are:

1. The tensile strength and stiffness properties;
2. The impact strength or fracture toughness.

This book is essentially about processes and a full analysis of the engineering properties of materials is not appropriate. The reader seeking an excellent introduction is recommended to ref. 2. Nevertheless, some description of the reinforcing effect of fibres is helpful at this point and this is given below.

12.3.1 *Strength and modulus*

To a good approximation, the tensile strength and Young's modulus of these composites follow a *law of mixtures* pattern. The effect is, as one would expect, highly anisotropic, depending on fibre orientation.

(a) Modulus

For a composite with the fibres laid in one direction, the full reinforcement is developed in the same direction. At an angle to the fibre lay direction the stiffness is less and it is at its minimum at right angles to the direction of lay. The law of mixtures which applies is

$$E_c = E_f V_f + E_m V_m$$

where E_c, E_f and E_m are the Young's moduli of the composite, fibre, and matrix respectively, and V_f and V_m are the volume fractions of the fibre and matrix, respectively.

In the angled direction the appropriate angled component of the fibre modulus applies. At right angles to the direction of lay there is no contribution from the linear stiffness of the fibres but only from their filling effect and the mixtures expression here is

$$E_c = E_m/V_m$$

Thus for a 50/50 blend $V_m = 0.5$ and $E_c = 2 \times E_m$. The stiffness at any angle lies between the two extremes and an envelope can be plotted to give a diagram, as shown in Fig. 12.2.

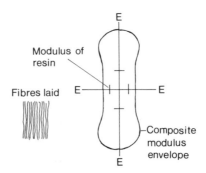

Fig. 12.2 Young's modulus distribution in a directionally laid composite.

(b) Strength

Tensile strength is even more anisotropic than modulus. A similar law of mixtures applies in the direction of lay, but at right angles there is no contribution at all from the fibres and there is only the resin strength available. The envelope (Fig. 12.3) is thus narrower at the cross direction.

(c) Cross laid fibres

If the fibres are laid in both directions, two overlaid envelopes can be drawn and a resultant pattern for the modulus or strength of the composite emerges

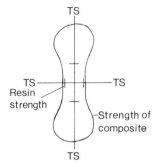

Fig. 12.3 Tensile strength distribution in a directionally laid composite.

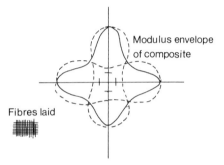

Fig. 12.4 Modulus pattern for a cross-laid composite.

(Fig. 12.4). If a random mat of fibres is used the strength and stiffness become isotropic.

The concept outlined above applies essentially to continuous fibres. When the aspect ratio (L/D) falls below 100, the modulus and strength enhancement decrease. At the level of fibre fills in e.g. thermoplastics, the modulus enhancement is about × 2 the matrix value, similar to the 'filler' effect in the cross direction with long fibres. At about 3 mm we reach the practical limit for reinforcement; shorter than this and the fibre becomes a particulate filler.

The enhancement of stiffness and tensile strength properties is also rather dependent on good adhesion between fibres and matrix. Weak bonds result in the fibres pulling out, rather than contributing to composite properties.

12.3.2 *Fracture toughness*

The fracture toughness of fibre reinforced composites is perhaps their most characteristic property. It is manifest in impact tests. A quite different situation holds, and the law of mixtures no longer describes the behaviour. For example [3], the work of fracture for an epoxy resin is in the range 100–300 J m^{-2}. That for the glass used to reinforce it is 8 J m^{-2}. The work of fracture of the composite is 40 000–100 000 J m^{-2}, a result that clearly does not follow the law of mixtures. The source of this very large increase in toughness is found in the bonding of the fibre and matrix. However, rather paradoxically, a weak bond is more effective than a strong one. If the bond is strong, a crack can propagate through the brittle fibre with little hindrance; if the bond is weak, debonding occurs and extra energy is needed to do the work of debonding. Also, the fibres fracture and their broken ends have to be pulled out as the fracture proceeds, which requires additional energy.

Thus we see that there is some conflict between the requirements for high modulus and tensile strength, which require strong bonds between matrix and fibre, and those for fracture toughness or impact resistance, which require weaker bonds. It is vital to specify correctly to obtain the desired result.

12.4 The hand lay-up process

12.4.1 *Process description*

This process is economically most suited to producing low quantities of large GRP mouldings, such as boat hulls and building panels. It is highly labour intensive. The essential operations are:

1. The mould is cleaned and a mould release agent applied. Often this can be a hard wax or a film of polyvinyl alcohol deposited from solution;
2. A gel coat of upe resin containing pigment (if required) and curing additives, is brushed evenly over the mould surface. This will form a pure resin outer surface to the moulding. Where the resin might drain down vertical surfaces a thixotropic additive may be used;
3. After the gel coat has become stiff, successive alternate layers of glass reinforcement, mat or cloth as required, and resin are applied, The glass layers are fully wetted and impregnated with the resin by rollers, or brushes used with a stippling action;
4. If required, a final resin-only sealing layer can be applied;
5. When the laminate has fully hardened it is stripped from the mould and trimmed to size, usually with a power saw.

12.4.2 *Features of hand lay-up*

1. The polyester resin hardens at room temperature without application of external heat;

Sheet moulding compound   227

2. The curing process does not evolve volatiles (c.f. phenolics and amino resins). This means the moulds are not pressurized and extremely large parts are readily made in a single moulding, with the mould open to the atmosphere;
3. Moulds can be made from cheap materials, because there is no pressure. Wood, GRP, plaster are used. Of course, this limits the number of units that can be made on them, and this must be taken into account in mould design;
4. Demould times are often long. This is necessary for large items to enable their construction before the resin cures. More than 30 min is common. If larger output volumes become necessary more than one mould may be needed and a large work area is required;
5. Critical mouldings, e.g. chemical storage tanks, may require a post-cure to develop optimum strength. Typical would be 3 h at 80 °C;
6. Thin areas in the moulding and sharp corners often become 'resin-rich', and contain no reinforcement. The properties are then deficient in these areas;
7. Only one surface is moulded, the other being rough;
8. The process is very operator-dependent, and a consistent resin–glass ratio is difficult to achieve. Considerable operator skill is needed to produce good mouldings;
9. The hand lay-up process is particularly useful for hand-building prototypes and mock-ups for other design routes.

Some of the features in the above list are attractions, whereas others are drawbacks. Once again, we see the need to compare the product specification with the potential offered by the process. Usually, there will be more than one viable solution to a design problem and the selection will depend on cost, previous experience and an element of personal preference.

12.5 Sheet moulding compound (SMC)

Sheet moulding compound is a prepared resin–fibre blend, often used as an alternative to the hand lay-up technique where longer, repetitive production runs are required.

12.5.1 *Preparation of SMC*

A sheet of polythene is coated with a layer of upe resin, blended with filler and curing additives. Chopped glass fibres are mechanically deposited on to the resin layer, and a second layer of resin paste is added. Another polythene sheet goes on top, and the whole sandwich is passed between rollers to impregnate the glass with resin and to consolidate. It is then wound into rolls.

228 Fibre reinforced plastics

Table 12.1 A typical formulation of SMC

Component	% by weight
upe	30.0
Peroxide initiator	0.5
Filler	38.5
Thermoplastic additive	6.0
Release agent	2.0
Pigment paste	2.0
Thickener	1.0
Glass fibre reinforcement	25.0
Total	100.0

The resin paste contains filler, catalyst, pigment, an internal mould release agent and a thickener. The thickener is usually magnesium hydroxide, which reacts with acid residues in the resin to form ionic bonds, and these convert the paste to a leathery consistency in about 36 h.

The initiator used is a high temperature type, e.g. PBT, described above. Sometimes thermoplastic additives (often low molecular weight polyethylene) are used to improve surface finish. A typical formulation is shown in Table 12.1.

12.5.2 *Moulding of SMC*

SMC is moulded by compression moulding techniques.

1. The required weight of SMC is cut from the roll. The polythene outer sheets are discarded.
2. It is placed in the mould to cover about 70% of the mould area, and of course correspondingly thicker: this gives the best flow pattern in the mould.
3. The press closes. The ionic thickening bonds are sheared by the heat and pressure. The temperature is in the range 140–170 °C, which activates the initiator. The resin flows to produce the moulded shape.
4. The press opens, allowing removal of the moulding, which may be trimmed during the next moulding cycle.

12.5.3 *Features of SMC process*

1. A moulding pressure of about 7.6 MPa is used, depending on viscosity and moulding temperature. This is much less than for injection moulding;

2. Cycle times vary, but are measured in minutes, perhaps over a range of 2–8 min, depending on part size, temperature, etc.;
3. High cost, steel moulds, sometimes chromium plated, are used;
4. This is a relatively high investment cost process, compared with hand lay-up, which necessitates long production runs, say in excess of 5000 per year.

12.6 Hand lay-up and SMC compared

12.6.1 *Advantages of SMC*

- both surfaces have moulded finish
- better consistency of composition and finish
- fewer finishing operations – moulding is more accurate
- higher output rate
- better potential for automation
- cleaner process – moulding material available in form ready for moulding
- no storage of resins, additives, glass, etc.

12.6.2 *Disadvantages of SMC*

- moulds are very expensive: for large parts, e.g. lorry cab panels, above £100 000 per tool
- moulding equipment is also capital intensive
- long production runs required to justify capital expenditure
- SMC itself has a 'shelf life' of 3–6 months at ambient temperature before it becomes unusable
- it is not usually practicable to include a pure resin gel coat
- There is the possibility of anisotropy in the properties of the moulded part. Placement of the sheets in the mould exerts a major influence, and overlaps are often required in corners and sharp curves to counteract movement during moulding. A problem is often flow of resin away from the glass to give resin-rich regions.

12.7 Dough moulding compound

Dough moulding compound (DMC) is another blend of glass and upe but uses short (3–12 mm) glass fibres. The resin, other filler, typically dolomite or ground limestone, together with other additives and initiator are mixed in a two-stage process. The liquid resin and small additives are stirred in a dip mixer of the Cowles dissolver type for about 20 min. The resultant blend is mixed with the filler and glass in a Z-blade mixer. A common way to

230   Fibre reinforced plastics

discharge the dough is through a screw located in the bottom of the Z-blade mixer. The DMC has a shelf life of about 7 days.

As we have seen in Chapter 9, DMC is commonly compression moulded, but there are also examples of its injection moulding (Chapter 8).

12.8 Process variants

There are several other variants of process using long-fibre reinforced resins. One has been cited earlier, viz. the compression moulding of carbon fibre reinforced PEEK. Two examples are described below, both giving highly orientated products.

12.8.1 *Pultrusion*

Figure 12.5 illustrates this process. A continuous strand roving is unwound from a reel to pass through a bath of resin in which it is impregnated with resin containing high temperature initiator. It is then drawn through a heated barrel which terminates in a shaped die, like that on an extruder. A shaped, cured profile emerges. Since the action is a drawing, or pulling one, the name 'pultrusion' has been devised for this process.

12.8.2 *Filament winding*

Figure 12.6 shows that this process also starts by impregnating a continuous strand of glass with resin. In filament winding the process continues by winding the strand on to a rotating former, usually of cylindrical shape. It is wound with any desired incident angle, and to various desired patterns. The products have great hoop strength, because the winding can cross at a varying angle to the equatorial. The process is used for storage tanks and pipes, where hoop strength is important.

12.9 Newer developments using thermosets

SMC and DMC are both characterized by supplying a prepared reinforced mix to the mould. One of the disadvantages is that the reinforcement may not be orientated in the most effective way; 'resin-rich' regions develop in thin sections and laying up blanks in the mould can be time consuming and not fully reproducible.

A series of processes has emerged in which the reinforcement is placed if the mould and the resin matrix is injected. These are called collectively *Liquid Composite Moulding*, LCM. At its simplest, this involves placing the

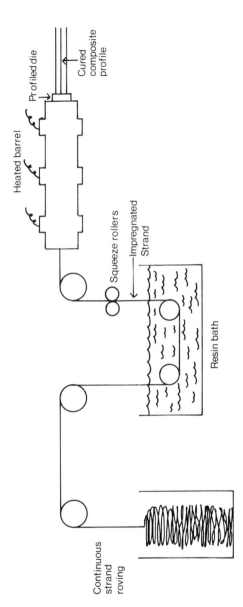

Fig. 12.5 Pultrusion process.

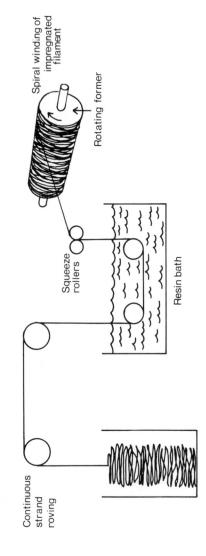

Fig. 12.6 Filament winding.

glass (usually) in the mould in a prescribed pattern, followed by injection of resin.

Manual placement of glass is slow and skill-dependent, and *preforms* are widely used. The glass is prepared to shape in a separate operation, and is lightly bonded with a thermoplastic binder, which softens when warm to accommodate to the mould. It is simply dropped into place at the moulding stage.

Variations also exist in the resin injection details:

- RTM – resin transfer moulding, uses premixed resin; rather like a development of hand lay-up with a closed mould;
- VARI – vacuum assisted resin injection, the vacuum helping to speed up the fill rate;
- SRIM – structural resin injection moulding uses a preplaced reinforcement or preform and injects a resin system which mixes in a mixing head on the way into the mould;
- RRIM – reinforced resin injection moulding mixes the resin on the way into the mould. A variant of Urethane RRIM. The glass is short and provides stiffening and increased heat distortion temperature rather than true reinforcement.

The main problem with LCM is the slowness and limited complexity of the resin injection, because of the resistance offered by the preform. A new process called *Network Injection Moulding,* NIM, has been invented by John Newton at Lancaster University, and his company 3D-Components (Cumbria). It allows fast injection of fast-curing resin systems into dense and complex reinforcement structures with 60% fibre content. It features low injection pressures (>7 bar) allowing cheap tooling. It will soon be offered extensively for automotive applications. It is already in use for inspection covers on garage forecourts, rated at 30 t loading.

12.10 Glass mat thermoplastics

We have mentioned the very high-tech Advanced Polymer Composite, APC from PEEK and C-fibre. This is far too expensive and high-tech for the automotive industry. Its place is in advanced aerospace applications where the dominant factor is *weight* at almost any cost. In automobile applications *unit cost* is dominant. Automobile economies reflect the acquisition cost of the vehicle; aerospace reflects operating costs,

e.g. $12 per kg for automobiles;
up to $650 per kg saved for the space industry.

Lower cost structural thermoplastics available are beginning to appear,

234 Fibre reinforced plastics

made from glass mat and thermoplastics, hence GMT. Quantity developments:

1983	325 t in automotive,	3 parts
1984	430 t	7 parts
1985	2000 t	20 parts
1986	3600 t	36 parts

Two principal preparative routes exist:

1. Hot, molten combination of the components in an oven followed by consolidation. The thermoplastic is extruded hot and the glass mat is laid on to it. There are obvious similarities in principle to SMC. These are the GEP/PPG Azdel and Azmet products, already mentioned in section 12.2.

    An alternative form for the glass mat is *polymat*, which is carded and needled and is binder free. It allows good penetration and high fibre content with good dimensional stability.
2. Wet processes based essentially on paper-making technology, in the STC process from Arjomouri, now acquired by Exxon. Glass and thermoplastic, often polypropylene, is dispersed wet and filtered, like a paper stock on a Foudrinier wire belt. This gives good dispersion of glass, breaking up bundles.

Equivalent performance: 40% glass bundles, 20% discrete glass fibres.

The result is a readily mouldable felt, with good flow properties.

Products to date include Peugeot 309 fan shroud, GM Astra van door components, air ducts, front bumper parts, door and seat structures under development.

The *radlite* process developed by Wiggins Teape, now acquired by GEP is another modified paper process. It uses aqueous foam to disperse resin powder and chopped glass, which is then spread on a porous belt with suction to collapse foam, after which the wet mat is dried.

Advantages offered include versatility in the final product form. The felt can be heated and pressed flat to give consolidated sheet; lighter treatment gives foam-like structure.

GE Plastics use the term *technopolymers* for both Azdel and Radlite types. TPS (technopolymer systems) is another term used.

Ahlstrom (Finland) is developing Reinforced Thermoplastic Composites, RTC. This is also based on the Wiggins-Teape Radlite process, but uses hot air ovens instead of IR re-heating, said to use only 10–20% energy of IR. Variants are:

- Alflow: easy-flow, 25% glass 65–70 bar moulding;
- Alstamp: high-strength, 80% glass, which can be in-mould coated with other polymers or fabric.

Moulding variants

These GMT processes will come on stream in the 1990s to deliver thousands of tonnes of GMT. They will be made by major suppliers rather than being in-house operations for moulders. They offer, compared with the metals they are expected to replace:

- equivalent strength
- lower weight
- faster cycling
- corrosion resistance
- low or no fabrication costs.

These are the first real contenders for *structural* plastics components in vehicles. The GEP Vector concept car will soon incorporate them as structural members.

12.11 Moulding variants

12.11.1 Press stamping

This is the most widely used process for GMT. A heated blank is positioned between matched mould halves, for compression moulding. Reduced pressure gives reduced densification (Radlite) and a more flexible product.

12.11.2 Variable densification

Multiple sheets are compressed together but not to the same extent throughout, to give variable density in the moulding.

12.11.3 Sandwich technique

This encapsulates a preshaped polyethylene or polyurethane foam component between STC skins.

Table 12.2 Liquid Composite Moulding, LCM

Name	Process
RTM	Resin transfer moulding. Glass or preform placed in mould. Premixed resin injected
VARI	Vacuum-assisted resin injection. Vacuum assists RTM process
SRIM	Structural resin injection moulding. Preform or glass placed in mould. Resin mixes in mixing head immediately prior to injection
RRIM	Reinforced resin injection moulding. Here the reinforcement – short glass – is incorporated with the resin in the mixing head
NIM	Network injection moulding. New development, rapid injection

Fibre reinforced plastics

Tables 12.2 and 12.3 summarize these new developments in both thermosetting and thermoplastic composites.

Table 12.3 Glass mat thermoplastics

Name	Process
AZDEL, AZMAT	Hot, molten combination of thermoplastic resin and glass mat
STC	Structural thermoplastic composite. Wet quasi-papermaking process
RADLITE	Foam based wet process
RTC	Reinforced thermoplastic composite. Radlite principle but hot air reheating

References

1. Brydson, J.A. (1984) *Plastics Materials*, 4th edn. Butterworths, London.
2. Powell, P.C. (1983) *Engineering With Polymers*. Chapman and Hall, London, Chs 4, 5, 6.
3. Weidmann, G. and Bush S. (1984) *Polymer Composites*, Unit 5 of *Polymer Engineering*. Open University, Milton Keynes, UK.

13
Rotational moulding and sintering

13.1 Evolution

Rotational moulding is a process for making hollow articles without seams. The only other such process available is blow moulding, discussed in Chapter 6.

Rotational moulding for use with polymers derives from an ancient process for casting hollow articles, originally in metals. A liquid, e.g. molten metal, was poured into a two-part mould, which was often rotated. The outside cooled and the mould was then inverted; the still-liquid core ran out, leaving a solidified skin, which was recovered by opening the mould. This process is still used, e.g. for toy soldiers. A later development was centrifugal casting to force-feed the liquid metal and for pipes, to hold the liquid against the walls, thus obviating the need for cores.

13.2 PVC slush moulding

The principle was extended for use with polymers first with PVC plastisols and a hot mould. The plastisol cures on the hot mould surface. The technique is to rotate the mould, with the plastisol in it, in two planes, to keep the plastisol evenly spread over the inner surface. The mould is usually heated in an oven.

Products manufactured by slush moulding of PVC include play balls, soft toys, and dolls' heads, bodies and limbs. Sometimes the product can be recovered through the filling aperture, but in other cases a split mould is needed, e.g. for balls.

A variation of the process is to cast on the outside of a heated former. Examples of products made in this way are gloves, handle grips and bellows gaiters.

13.3 Powdered polymers

Powders of thermoplastics can be sintered and finally fused to form coatings.

13.3.1 *Rotational moulding of powders*

The technique is the same as that described above for slush moulding. A measured quantity of polyethylene powder, usually high MFI for easy flux and flow, is placed in the mould which is then placed in an oven and rotated in two planes to give a tumbling action, which sinters the PE at the walls. Some further details are given below.

(a) *Materials*

The polymer mainly used is polyethylene. There is some claim for polypropylene, ABS, nylon and polycarbonate for vandal-proof transparent lighting globes. The limitation is the risk of polymer degradation over the necessarily long cycle times. In some cases this problem can be alleviated by flushing with nitrogen to prevent oxidation.

Some use of foamed polyethylene is reported, using a two-shot process:
(a) shot 1: skin formation;
(b) shot 2: contains a blowing agent (azodi-carbonamide).

A recent development is the use of thermoplastic elastomers, for such applications as children's bicycle wheels and tyres, which are moulded integrally.

(b) *Heating*

The usual heating method is to use a forced-convection air oven, gas or oil fired. Temperatures of 190–400 °C are required. Other methods that have been used are molten salt baths or sprays. For example, a mixture of potassium nitrate, sodium nitrate and sodium nitrite gives the required temperature regime, but these mixtures are unpleasant to use, being oxidizing and corrosive. For simple moulded shapes, hot oil jacketed moulds are a good alternative to the oven.

(c) *Cycle times*

Cycle times are long because heating and cooling are required, usually by convection methods, and there is only one metal surface available for heat transfer, which is therefore slow. For example a 1 m diameter mould would be 2.5 cm thick in aluminium or 0.5 cm thick in steel. Cycle times in a hot air oven for different coating thicknesses are:

0.13 cm coating: 10 min
0.25 cm coating: 12 min
0.40 cm coating: 14 min.

(d) *Products*

Rotational moulding is a method well suited to the manufacture of hollow goods such as footballs, marine buoys, surfboards (later PU foam filled), road signs, traffic cones, bins and containers, petrol tanks. The method is capable of more complex shapes than are possible by blow moulding.

Products are characterized by:

- zero orientation; there is no shearing at all during processing, and dimensional stability is therefore excellent;
- very high physical properties – equal to those published in suppliers' technical service notes, which most processes do not achieve. Rotationally moulded bins and traffic cones are exceptionally tough;
- uniformity of wall thickness, compared with blow moulding or thermoforming, especially in corners.

Tooling costs are very low; the forces involved in rotational moulding are small, and cheap moulds are satisfactory. The plant is also cheap by comparison with other moulding processes.

13.3.2 Surface coating

Tough, resistant and decorative coatings can be applied to steel articles by sinter coating. Examples are the handles of tools, mowers, etc. The best technique uses a fluidized bed of the polymer powder. Air is passed at a suitable rate from a large number of ports vertically upwards through a bed of powder. The powder now behaves like a liquid. If the article to be coated is heated and dipped into the fluidized bed, a layer of polymer sinters on to it and adheres. Subsequent oven treatment completes the fusion of the coating.

A variant is the external coating of steel pipe with polyethylene as a protective coating. The coating offers direct protection; it also helps when the pipe is to be protected by cathodic protection, by providing a high dielectric coating which minimizes current consumption. The process starts by heating the pipe to 290 °C; this ensures good adhesion of the polyethylene by oxidizing its surface upon contact. Uniform heating is essential for even coating, and it is achieved by rotating the pipe against a row of gas flames along its length. The polyethylene is applied by strew coating from above the rotating pipe until a thickness of 2.5–3.5 mm has built up, which takes 2–7 min. The maximum practicable coating thickness varies with the pipe wall thickness, and is approximately $0.4 \times$ wall thickness. When coating is complete, the pipe is transferred to a holding area for fusion of the polymer to continue spontaneously, using heat already in the pipe which is insulated by the polymer layer. Finally, the pipe is cooled by blowing cold air through it.

240 Rotational moulding and sintering

13.4 Comparison of rotational and injection moulding

There are many products which could equally be made by rotational moulding or by injection moulding. For example, petrol tanks can be made in one piece rotationally, with exceptionally good property development, but virtually always such items are made by injection moulding two halves and welding them together. Why is this so?

The answer lies in the respective economics of the processes. There is an example by Morton-Jones and Lewis [1] of a comparative costing on a polypropylene battery box made by both methods.

In this study the Fourier method (Chapter 2) is used to estimate the cooling times by both methods, using the following parameters for a square battery box 10 cm sides and 4 cm thick base (Table 13.1). For injection

Table 13.1 A comparison of rotational and injection moulding

Injection moulding	*Rotational moulding*
Melt temperature 250 °C Heat distortion temperature 115 °C Mould temperature 60 °C	Mould temperature 300 °C

moulding the Fourier method finds the cooling time, using the half thickness and cooling from both sides.

Cooling time by injection moulding = 33 s
Assume this is 60% of total cycle, giving a cycle time of 55 s.

For rotational moulding, the Fourier calculation is used for heating and cooling. Only one side is available for heat transfer, and full thickness must therefore be used.

Heating time in rotational moulding = 170 s
Cooling time = 105 s
Total cycle time = 275 s = 4 min 35 s.

The material costs are the same for both methods so that the process costs represent the comparison. These are in the ratio 275:55 (i.e. 5:1).

The missing factor is the tooling cost. An injection tool would cost probably tens of thousands of pounds, which has to be amortized in the product cost, and requires thousands of mouldings to warrant its cost. Thus we see that injection moulding gives a much cheaper product, if thousands are required. Rotational moulding would be preferred for a few hundred products, which could not justify the cost of injection tooling.

For this product, injection would be preferred, because a product such as a battery box, or a petrol tank, would be sold in thousands or not at all.

Rotational moulding scores when smaller numbers are required, e.g. surf boards, traffic cones, until the market begins to be large enough for injection moulding to enter. Then the business tends to go to the lower processing cost injection moulding.

Reference

1. Morton-Jones, D.H. and Lewis, P. (1984) *Polymer Processing*, Unit 3 of *Polymer Engineering*. Open University, Milton Keynes.

14

PVC and plastisols

14.1 Introduction

Plastisols are paint-like mixtures of PVC powder and plasticizer. An independent area of technology exists to exploit them. As we have noted in previous chapters, the polymer processing industry contains a number of these discrete technological areas, within which features peculiar to the special materials are exploited. Other examples, seen throughout this book, include:

fibre reinforced plastics
PVC calendering
thermoset compression moulding
rubber processing – an industry in its own right
the thermoplastic processors – injection moulding and extrusion.

14.2 Polyvinyl chloride (PVC)

PVC is commercially available in a number of different types, which are displayed in Fig. 14.1. The characteristics of these various grades are

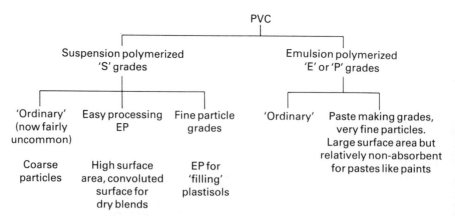

Fig. 14.1 Types of PVC.

Polyvinyl chloride

examined in more detail below. The main division is into *suspension* and *emulsion* types. The names refer to the polymerization process used in each case. The mechanisms of the suspension and emulsion polymerization routes are different and the polymer types which result reflect their different origins.

14.2.1 Suspension polymers

These are divided into the 'ordinary' grades, with particle size of 50–100 μm and the 'easy processing' or EP grades of similar particle size as the ordinary type but a surface area : volume ratio eight times that of the ordinary grade. This is achieved by selecting a suitable suspension dispersing system, e.g. maleic acid–vinyl acetate copolymer, which gives an irregular particle shape, with many voids and cavities.

The EP grades are used for dry blends. There is a description of the production of PVC dry blends, made from EP grades, in Chapter 3. PVC calendering, which uses suspension polymer, is discussed in Chapter 10.

The now largely outdated 'ordinary' grades were made using, e.g., gelatin as a protective colloid, and have smooth spherical particles.

14.2.2 Emulsion polymers

Again we find an old 'ordinary' grade not much used today. This, in common with all types of emulsion polymer, has a much smaller particle size (0.1–1.0 μm). These polymers were originally designated with an 'H' prefix, e.g. the ICI *Corvic HO*, now obsolete. The Norsk Hydro polymer *Norvinyl G*, and the Huls polymer *Vestolit GH* were similar. The particulate structure of these polymers was a sort of hollow-centred sphere, called a 'cenosphere'. They have a high surface area to volume ratio, and rapidly absorb plasticizer to give high viscosity pastes, which usually require refining on a three-roll mill before spreading. These original grades of emulsion polymer have been largely superseded by the modern 'paste' or 'stir-in' grades, which are not cenospheres. Their particle size range is 0.2–1.5 μm, and pastes are readily made from them simply by direct mixing without the need for refining.

14.2.3 Particle size distribution

Particle size distribution is important as well as average size. The average size of the polymer particles determines whether the paste will be stable, i.e. non-settling, and will influence the appearance and properties of the finished product. The size distribution controls the plastisol viscosity.

Wider size distribution allows more efficient packing and hence lower paste viscosity. The use of 'filler' polymers, which are suspension polymers

244 PVC and plastisols

of relatively fine particle size, helps to broaden the overall distribution and lowers viscosity. They can be used in ratios of up to 25% of the total polymer.

Some polymers give *dilatant* pastes, usually because of irregularly shaped particles which 'log jam' when sheared; often the effect is most pronounced when the size distribution is narrow. Partial replacement with filler polymer usually overcomes the problem.

K-Value

PVC polymers are also graded by K-value, which is a measure of molecular weight, usually found from measurement of intrinsic viscosity in cyclohexanone solution. Note that the 'K' is not the 'K' in the Mark–Houwink equation (see Chapter 1). It is obtained from the Fikentshner equation and is often called the Fikentshner K-value. It has a value of 100 K in this equation

$$\frac{\log_{10} \eta_{rel}}{C} = \frac{75K^2}{1 + 1.5KC} + K.$$

It usefully grades PVC polymers by their solution viscosity and hence their molecular weight over a range of about 35–80.

A PVC of K-value 35 has low molecular weight and will be very easy to process, but will not deliver outstanding properties. At K-value 80, processing will be quite difficult but if it can be done well the properties of the product will be exceptional. It is often better to select a medium K-value and process to full development of properties than to attempt to use a high K-value polymer, but to process it slightly inadequately.

14.3 Plasticizers

Plasticizers used in PVC fall into several classes. The most widely used are high boiling point esters, e.g.

  di-octyl phthalate (DOP)
  di-iso-octyl phthalate (DiOP)
  di-alphanyl phthalate (DAP). DAP is the phthalate ester of a mixture of C_7–C_9 alcohols; it thus approximates to DOP
  tri-xylenyl phosphate (TXP), which gives fire protection
  adipates (C_6) and sebacates (C_{10}) of C_{10}–C_{12} alcohols give low temperature flexibility, because of their low T_g.

Polymeric plasticizers, e.g. polyethylene glycol phthalate, are more resistant to leaching out by solvents or washing detergents, but they are little used in plastisols because of their relatively high viscosity. They find application in calendered goods for tarpaulins and rainwear.

Secondary plasticizers, e.g. chlorparaffins, (e.g. the *Cereclor* range from ICI) are useful for reducing cost. They cannot be used alone; they are not quite compatible with PVC but are effective as diluents. They count as plasticizers when formulations are being worked out. Usually they would be used to replace between a quarter and one-half of the primary plasticizer.

Extenders are hardly used at all today. They are incompatible mineral oils which were fairly extensively used in the past, but they are unsatisfactory and bleed out later. Old electrical installations sometimes betray their use as a sticky exudate.

14.4 Fillers

Fillers can be used up to about 30 pphr, but PVC cannot be loaded to the same extent as rubber. Fillers in plastisols rapidly shorten and stouten the paste and are not used in high proportions, although a small addition can give a good surface for decorative printing.

14.5 Stabilizers

Several classes of chemical compounds are used as stabilizers. Without them, PVC quickly degrades during processing, turning dark coloured and losing its properties. It evolves hydrogen chloride which catalyses the degradation, accelerating further degradation. It does not depolymerize to vinyl chloride monomer (VCM).

1. The first stabilizer was white lead (basic lead carbonate). This is not by modern standards an especially good stabilizer, and the high temperatures in rigid PVC extrusion cause carbon dioxide to be evolved, which leads to porosity.
2. Dibasic lead phosphite is used in some products for outdoor applications, e.g. some window frames, where good light stability is essential, but its use is excluded in other applications because of its poisonous nature. Other stabilizers offering light stability are tin ester, e.g. dibutyl tin maleate and newer products developed by Akzo Chemie.
3. Various metal soaps are extensively used, in blends which act synergistically. A blend of barium/cadmium/zinc was widely used until recently but the toxic nature of cadmium has led to its being discontinued in many countries. Blends using calcium are used instead although they are somewhat less effective.
4. For processes with a low heat experience, calcium stearate is effective and non-toxic, but soon becomes ineffective if processing is extended.
5. Thio-tin salts and esters give good colour retention, and high temperature resistance for short periods.

246 PVC and plastisols

6. Organic phosphites also give good colour retention and enhance the performance of other stabilizers.
7. Epoxidized vegetable oil derivatives are often used as 5–10% replacement of plasticizer, when they act as stabilizer synergizers.

For most applications a 'cocktail' containing two or three types is used to ensure tolerance of high temperature and also a long processing life. One of the preoccupations of the PVC formulator is to ensure that the product meets the requirements of toxicity regulations, especially if the product is to be used in food packaging, e.g. drinks bottles by blow moulding, or cling film by calendering or plastisol route.

14.6 Blowing agents

Plastisols are frequently foamed, e.g. for soft expanded leathercloth and cushioned vinyl floor and wall coverings. The most widely used blowing agent is azodicarbonamide AZDC

$$\underset{H_2N}{\overset{O}{\diagdown}}C-N=N-\underset{NH_2}{\overset{O}{\diagup}}C .$$

A typical formulation for a foamed plastisol would be:

Paste polymer,	100 p.b.w.
DAP,	80 p.b.w.
Epoxidized oil,	5–10 p.b.w.
Filler/pigment,	0–10 p.b.w.
AZDC,	2–4 p.b.w.
Cd/Zn or Zn or Ba/Zn soap,	1.5–2.5 p.b.w.

14.7 Substrates

Plastisols are spread on to a supporting base material or 'substrate', which serves to carry the liquid in the early stages and then becomes part of the product.

Leathercloths, for upholstery, luggage etc., use woven or knitted cloths in cotton, nylon or polyester. Care must be taken to ensure that the selected fabric will tolerate processing conditions, especially oven temperatures. Some leathercloths are made upside down; the plastisol is coated on to release paper and the fabric is laid into the coating. Wallcoverings are usually coated on to paper.

Cushioned floorcoverings have used asbestos felt in the past, but this is now replaced with a fabric-supported PVC base. Other types of heavier duty floorcovering have in the past used jute hessian and needlefelt, and cork composition.

14.8 Formulation

The softness or stiffness of the eventual product will be mainly decided by the PVC/plasticizer ratio. These two components together make up the 'binder' and are regarded as the total resin or rubber. Thus, when we speak of additions in pphr the 'hundred' of rubber or resin is the combined PVC and plasticizer.

The characteristics of some typical ratios are given in Table 14.1.

Table 14.1 The characteristics of some typical PVC/plasticizer ratios

PVC	Plasticizer	Description
80	20	Very stiff paste and product
70	30	Stiff
60	40	Soft product
50	50	Very soft

At low levels of plasticizer high paste viscosity becomes a problem. Different makes and grades of polymer offer different characteristics. Some examples are:

Breon P 130/1, dilatant paste even at 65/35
Corvic P 75/578, low viscosity; Newtonian or slightly pseudoplastic response
Vestolit E 8001, medium viscosity; pseudoplastic
Pevikon PE 709, high viscosity, but pseudoplastic. 60/40 limit in use.

The high viscosity polymers are not necessarily a disadvantage; they are valuable for soft, highly plasticized products, to give some processing structure to the paste. Thus even a dilatant response, although undesirable in some cases, will allow a thick coating to be applied and then, when the shear is removed, flow to a superlative glossy finish. As we have already seen, control of viscosity is available through judicious use of filler polymers. Filler polymers tend to give a more matt finish and this in itself may sometimes be desirable.

14.9 Processing

14.9.1 *Mixing*

Plastisols are mixed either in dip mixers (Chapter 3) or in speciality mixers which are like giant food mixers; in fact food mixers are frequently used on

248   PVC and plastisols

the laboratory scale for experimental mixes. These mixers may be equipped with vacuum facilities for mixing under vacuum, to exclude trapped air.

The feature of the mixing is that it is low shear, compared with rubbers and hot PVC. It is, in fact, distributive mixing, with no attempt at dispersion. The resultant paste is paint-like in nature – in fact pastes are often called paints in the trade. It has a limited pot-life; the plasticizer gradually swells the polymer and after a day or two the viscosity is obviously increased.

14.9.2 Coating

Two principal methods of coating are used: knife coating and reverse roll coating.

(a) Knife coating

This is shown in Fig. 14.2. The plastisol is held in a small bank behind the knife as the substrate passes over the roller. A 'shoe' on the downstream side of the blade prevents tailing or streaking of the coating. This method of coating gives a smooth surface and a product of uniform thickness, even with an uneven substrate.

Fig. 14.2 Knife-on-roller coating.

(b) Reverse roll coating

Figure 14.3 shows the reverse roll coating process. Its principle is to form a film between the metering and casting rolls and then to transfer it to the substrate as it passes through the nip formed by the casting roll and the bottom or carrying roll. At the coating point, the casting roll and the substrate are travelling in opposite directions, hence the term 'reverse roll coating'. The characteristic of this method of coating is that it delivers a coating of uniform thickness; any unevenness in the substrate is repeated on the coated surface. The even coating thickness is important in foaming applications, to ensure even blown thickness with blow ratios as high as 3 or 4 to 1 in some cases. The shear rate is very high as the substrate wipes off the coating from the casting roll, a typical value being 3×10^3. Any tendency towards dilatancy in the paste can have dramatic results; the stress developed as the apparent viscosity increases under shear can be sufficient

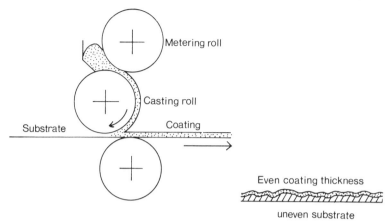

Fig. 14.3 Reverse roll coating.

to stall the machine. At slightly lower levels it causes a chattering or juddering as the stress rises. The process stops, the shear therefore falls again and the process restarts, only to go through the same cycle again, in a stop-start sequence. Figure 14.3 shows the action of a reverse roll coater.

14.9.3 Curing

The cure of a PVC plastisol is effected by heating. The coated material passes directly through an oven where the PVC–plasticizer blend fuses. The cure is essentially physical; the PVC and plasticizer have been converted from a paint to a molten rubber in the oven. The cross-linking of this rubber is through microcrystalline domains which are thermally labile. Plasticized PVC may thus be regarded as the first of the thermoplastic rubbers.

The oven has a series of graded temperature zones. One of the problems in using emulsion polymer is that there are usually traces of emulsifiers still present; these are hygroscopic, with the result that the paste contains traces of moisture. If the coated material is plunged directly into an oven at curing temperature, the moisture flashes off to form blisters. Grading the oven temperature allows the moisture to escape more slowly and harmlessly before curing temperature is reached. A typical oven would be 20–30 m long, zoned at 110, 180, 220 °C. Ovens have forced convection permitting control to ±1 °C across the width. There is a conveyor, usually of woven wire, running through the oven to carry the coated substrate. Figure 14.4 illustrates the plant. As the material emerges from the oven it may be passed through a pair of rollers to impart a smooth or embossed finish, e.g. a leather grain or geometric texture. Alternatively, an emboss can be applied later by reheating with IR heaters before passing through an embossing nip.

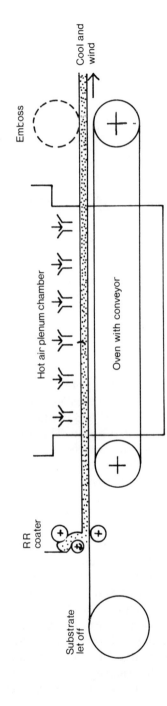

Fig. 14.4 Coating and oven curing.

14.10 Chemical embossing

The technique of chemical embossing was invented by Congoleum Nairn of the USA in the 1950s. It provided a method for producing a decorative printed pattern and a surface texture perfectly in register with one another. This is extremely difficult to achieve by mechanical means, because the material always stretches to some extent during coating and curing, sufficiently for a printed pattern and a final emboss to lose register. A general embossed grain can be applied, but not a texture that follows the printed pattern lines.

Chemical embossing (also known as *debossing*) uses a foam to give the texture. The procedure is as follows

1. A substrate is coated with a plastisol which contains AZDC as blowing agent and also a *kicker*, usually zinc oxide, which lowers the decomposition temperature of the AZDC from 230 °C to about 190 °C. Reverse roll coating is preferred to ensure an even coating thickness;
2. The material is passed through an oven at about 110 °C, which 'sets' the plastisol. It becomes cheesy in consistency, solid and dry but without strength;
3. A decorative pattern is printed on the set surface. Selected colours of the printing ink contain an inhibitor which reacts with the zinc oxide kicker to deactivate it, but only where the special ink has been printed. Typical inhibitors are fumaric acid and trimellitic anhydride (TMA), both acidic materials capable of forming complexes with the zinc oxide. Printing methods employed are gravure, using as inks solutions of PVC, or rotary screen print, using plastisol inks. Gravure techniques were used exclusively in the early days, using trichromatic colouring for the very sophisticated patterns demanded at that time;
4. A final coat of plastisol is coated over the print. After curing, this will act as a clear wear layer to protect the print in service;
5. The final process is a pass through the curing oven at about 200 °C. In this oven, the foam coating fuses and blows, except where the inhibitor is located; the top coating fuses to become a clear wear layer, and it follows the contours of the foam layer beneath it. The result is a product carrying a pattern with texture induced chemically by the print itself, e.g. mortar lines in a printed tile design.

The whole process is shown diagrammatically in Fig. 14.5.

The chemical embossing process was originally used for floorcoverings, under the name *Cushionflor*. It was redeveloped under licence in the UK by the Nairn–Williamson company, and later taken up by a number of other companies. At one time it was the subject of acrimonious patent litigation in the USA. More recently, the process has been adapted to the production of

252 PVC and plastisols

```
─────────────────────────────── Substrate

═══════════════════════════════ Coated with foam plastisol
                                set at 110°

≡≡≡≡≡≡≡≡≡≡≡≡≡≡≡≡≡≡≡≡≡≡≡≡≡≡≡≡≡≡≡ Print; ink contains inhibitor

≡≡≡≡≡≡≡≡≡≡≡≡≡≡≡≡≡≡≡≡≡≡≡≡≡≡≡≡≡≡≡ Coat with wear layer

∿∿∿∿∿∿∿∿∿∿∿∿∿∿∿∿∿∿∿∿∿∿∿∿∿∿∿∿∿∿∿ Cure and blow. Foam
∘∙∘∙∘∙∘∙∘∙∘∙∘∙∘∙∘∙∘∙∘∙∘∙∘∙∘∙∘∙∘ Inhibited at inhibitor ink
```

Fig. 14.5 The chemical embossing process.

wallcoverings, where it offers outstanding design potential, as demonstrated in the range of *Kingfisher* wallcoverings manufactured by Nairn Coated Products Ltd.

14.11 The plastigel process

The final section of this book is somewhat anecdotal, in that it describes a process unused now for many years. However, its story is worth telling for two reasons; firstly it places on record the details of an extraordinary process, and secondly it illustrates the value of lateral thinking, and the way in which experiences in different fields can benefit one another.

The plastigel process was developed at the company of Jas. Williamson & Son Ltd. of Lancaster (now absorbed into Nairn Floors) during the 1950s. Williamson's were premier manufacturers of linoleum and also pioneer developers of plastisol technology, for leathercloths. The plastigel process was devised to manufacture the first PVC floorcovering, under the name *Crestaline*. In effect, the process combined linoleum technology and plastisol technology. Linoleum is produced by calendering a composition based on linseed oil, which is mixed in Banbury mixers. To provide shear, the oil is oxidized to a stiff gel, called *cement*. The resultant composition then requires a cold calender to avoid smearing its characteristic marbled decoration. The problem was how to make a PVC-based composition which would process on the existing linoleum machinery.

The ingenious solution was to make a PVC equivalent to the linoleum

The plastigel process

cement. This was done by dissolving emulsion-grade PVC in plasticizer at a concentration of 10%. The result is a stiff jelly, very sticky, but offering the correct shear properties for linoleum-style mixing in a Banbury mixer. A composition was developed which contained further PVC, whiting as filler, pigment and calcium stearate stabilizer; all the required plasticizer was in the form of the new 'cement'.

The formulation went through several developments. The final version, which ran for some years very successfully was:

Plasticizer (a 1 : 1 blend of DAP and TXP),	0.85
PVC: emulsion polymer (*Vestolit GH*),	1.0
Filler polymer (*Corvic* D55/3),	0.75
Whiting and pigment,	1.25
Calcium stearate,	0.05.

Some of the emulsion polymer was taken to make the 10% cement. The main mix was then made in a Banbury mixer, water cooled throughout, until a plasticine-like composition was ready. This was then granulated through a series of breakers to the size of marbles. Different colours of granules were blended for feeding to the calender. Note that this process used the old style of emulsion polymer. Only this type gave controllable mixing, without overheating.

The calendering process was virtually identical with that used for linoleum; it is quite different from the process familiar to rubber and hot PVC technologists. It uses cold calenders (0 °C face roll). Two two-roll machines are used; the blended granules pass through the first machine, which streaks the colours linearly; this streaked sheet is cut into planks, which are stacked in a staggered arrangement and presented to the second calender so that the streaks run crosswise. The second calender elongates the colours at right angles to the first, to give a marbled effect. The marbled, calendered sheet is pressed on to a hessian backing in a machine resembling a rotocure (Chapter 11).

The PVC product (*Crestaline*) was cured by passing it through a curing oven, exactly like any conventional leathercloth product.

The Crestaline process may be regarded as a plastisol process in which the plastisol is so stout that it requires a calender to spread it. It arose from the coincidence in one company of two quite different technologies, and it illustrates how they were able to cross-fertilize to evolve an almost unique process. 'Almost', because another company, in Stuttgart, also processed linoleum and plastisol PVC: They also, quite independently, developed a similar, if not quite identical, process.

The *Crestaline* product was very successful, technically and commercially. It ran for many years, and thousands of yards of it are still giving excellent service.

Index

Ablative shield 177
Acetal *see* Polyacetal
Acrylic *see* Polymethyl methacrylate
Acrylonitrile-butadiene-styrene, ABS 51–3, 144, 166
Additives 55–8
  chemical additives 56
  modifying additives 55–7
  protective additives 57–8
  see also Carbon black
Antioxidants 57, 198
Antiozonants 57
Antistatic agents 58
Apparent viscosity, *see* Viscosity
Arnitel 218
Auma vulcanizing press 211
Azdel, Azmet 223, 234, 236

Ball mill 63
Banbury mixer 66, 208, 252–3
Baths, domestic 144
Bingham body 38
Blow moulding 126–37
  blow pin 127, 133
  blowing 128
  blowing pressure 129
  bottles for carbonated drinks 133–7
  extrusion blow moulding, EBM 126–33
  injection blow moulding, IBM 133–7
  injection moulding in IBM 135
  large blow mouldings 133
  parison, parison sag 126–8
  pinch-off 128
  preform 126, 133
  product properties 131
  stretch-blow 133, 136–7
  wall thickness 129, 135, 136

Blowing agents 56, 169, 238, 246, 251
Blown film 118
  freeze or frost line 119
  polypropylene 120
  stabilization of bubble 119
Boats 144
Boil-in-the-bag 137
Brabender Plasti-corder 215

Calendering 35, 37, 186
  cold, plastigel process 253
  contour grinding 188
  crossed axes 187–8
  orientation 188
  PVC calenders 188
  roll bending 187–8
  rubber calenders 188
  separating forces 187
Carbon black 202–6
  acetylene process 203
  bound rubber 205
  channel process 203
  classification 203–5
  DBPA test 202, 205
  effect on compounds 205–6
  furnace process 203
  iodine adsorption 202
  manufacture 202–3
  reinforcement mechanism 205
  structure 202, 205
  surface area 202, 205
  thermal process 203
Cavity transfer mixer, *see* Extruders
Chain scission 57
Cinpres 170
Cola drinks 133, 134
  see also Blow moulding
Compression moulding 35, 176–83

Index   255

advantages 182
compounds 179
moulds 187
moulding pressures 180
pot life 179
presses 180
process 179–80
resins 176–9
rubbers 211
volatiles 179
see also Transfer moulding
Consistency Index 39
Copolymers 8
 alternating 9
 block 8
 graft 9
 random 8
 terpolymers 8
Cowles dissolver 64, 229
Creep 17–22, 135
 creep modulus 18, 39
 creep rupture 20
 viscoelastic interpretation 21
Cross-linking 8, 56, 186
Crystallinity 10, 15, 117, 132, 135, 165
 stress induced 136–7, 165
Curing 48
Curometer (Shawbury) 215
Cushionflor 251
 see also Plastisols

Deborah Number 107–9, 117, 128, 143
Dies, extrusion 105–11
 bambooing 110
 characteristic 102–5
 coathanger dies 122, 210
 design 112
 die entry effects 109
 die swell 110
 fishtail dies 122
 flow patterns 105
 land 113
 orange peel 110
 sharkskin 109
 sheet dies 121
 T-dies 122
Dilatancy 38
Dip mixer 65, 247–8

Dough moulding compound, DMC 171, 173, 174, 179, 229
Drawdown 114
 see also Extrusion-based processes

Elastomers, *see* Rubbers
Electrical cable insulation 115
Electrical fittings 177, 178
Environmental stress cracking, ESC 144
EPDM, *see* Rubbers
Epoxy resin 48
Extenders 56
Extruders
 breaker plate 77, 78–80
 cavity transfer mixer 80
 characteristic 102–5
 compression zone 76
 die, *see* Dies, extrusion
 feed zone 76
 flight 74
 metering zone 77
 mixing head 80
 nylon screw 76
 operating point 102
 polyethylene screw 76
 PVC screw 77
 roller die 210
 rubber 116, 210
 screen pack 77
 screws, *see* Extruder screws
 single screw 74–97
 speciality features 80–2
 twin screw 97–102
  see also Twin screw extruders
 venting 81
 zones 76–8
Extruder screws 92–7
 barrier flight 92
 helix angle 94–6
 Maillefer 92
 melting efficiency 92–3
Extrusion 35, 74–111
 adiabatic 84
 analysis of flow 84–9
 co-extrusion 124, 171
 conveying 83
 drag flow 83, 84–6
 flow mechanisms 82–4

256 Index

Extrusion *contd*
  heating and cooling 83
  influence of polymer properties 89–92
  isothermal 84
  leak flow 83, 88–9
  melting 82
  pressure flow 83, 86–8
  rubber 116
  total flow 89
  turning memory 78–80
Extrusion-based processes 112–25
  blown film 118–21
    see also Blown film
  cast film 122
  cross-head extrusion 115
  drawdown 114
  extrusion coating 122
  fibres, *see* Synthetic fibres
  netting 124
  pipe extrusion 112–14
  profile extrusion 112–15
  sizing mandrels 114–15

Fibre reinforced plastics 220–36
  carbon fibre 220
  dough moulding compound, DMC 229
    see also Dough moulding compound
  fibres 220
  filament winding 230
  fracture toughness 225
  glass fibre 220
  glass mat thermoplastics, *see* Glass mat thermoplastics
  hand lay-up 226–7, 229
  Kevlar 220, 221
  liquid composite mould, *see* Liquid composite moulding
  modulus 221, 224
  processes 226–30
  pultrusion 230
  resins 221–3
  sheet moulding compound, SMC 227–9
  strength properties 223–6
Filament winding, *see* Fibre reinforced plastics

Fillers 201–7
  calcium carbonate 207
  china clay 206–7
  mineral fillers 206–7
  silica 207
  see also Carbon black
Fire hose 117, 212
Flow behaviour index 39, 95
Flow in channels 41–4
  driving pressure 43
Foam-cored moulding, *see* Structural foam and Sandwich moulding
Formica 178
Fourier number 50–3, 240
Free radicals 7, 48
Freezer liners 144

Garage doors 144
Glass mat thermoplastics, GMT 233–6
  Advanced Polymer Composite, APC 233
  Azdel, Azmet 223, 234, 236
  Radlite process 234, 236
  Reinforced Thermoplastic Composites, RTC 234, 236
  Structural Thermoplastics Composite, STC 234, 236
  technopolymers 234
Glass reinforced plastics, GRP *see* Fibre reinforced plastics
Glass transition 13, 14–15, 49, 133, 136, 143

Heat setting 137
High speed (Henschel) mixer 62
Hollow shapes 126, 133, 237
Hoop stress 21, 130, 165
Hydrogen bonds 14, 186
Hysteresis 191–2
Hytrel 218

Impact modifiers 56
Injection moulding 35, 37, 44, 146–75
  basic process 146–9
  'burning' 155
  cavity or impression 147, 152
  channels 147
  clamp force 151, 170
  clamp unit 146, 151

Index

co-injection 170
computer-aided design, CAD 162–4
cooling of mouldings 50–3
cooling of moulds 53, 155
design aspects 161–4
effects of heat and pressure 164
ejector pins 147
family moulds 162–4
foam-cored, *see* Structural foam and Sandwich moulding
gates, gating 147, 155–60, 162
hold-on pressure 147, 160, 161, 167–8, 170
injection 147, 160, 167–8
injection pressure 152, 160, 167–8
machine 149–56
material response 160
microprocessor control 168
mould or tool 147, 152–6
mould, three plate 152–4
mould, two plate 152–4
nozzle 150
orientation 165–6
polymer selection, EPOS 164
product quality 160–8
projected area of mouldings 151, 170
regrind 168
runners 147, 152–4, 162, 169
shot size 152
shrinkage 166
sink marks 161, 170
sprue 147, 152
sprueless moulding 168–9
stress concentrations 162
thermosets 173–4
venting 154
voids 161
weld lines 161
Intermix 67
Internal mixers 66
Isochronous curves 18
Isometric curves 18
Isotactic, *see* Polymers

Kevlar 220, 221
   see also Fibre reinforced plastics

Latent heat of fusion 45, 47

Latex 195
Liquid Composite Moulding, LCM 230–3
   Network Injection Moulding, NIM 233
   reinforced resin injection moulding, RRIM, 233
   resin transfer moulding, RTM 233
   structural resin injection moulding, SRIM 233
   vacuum assisted resin injection, VARI 233
Lubricants, processing 58

Mark–Houwink equation 6
Masterbatch 205
Melamine-formaldehyde 48, 173, 178
Melinex 11
Melt Flow Index, MFI 44–5
Melting of polymers 45–7
Mixing 53–73
   computer controlled 209
   extensive, blending or distributive 59–61, 61–4
   forces in mixing 71
   intensive, compounding or dispersive 59–61, 64–9, 208
   procedure 68–9
   processes and machines 61–71
   routes 71
   summary 70
   thermoplastics 70
   turbulence 72
Modulus 18, 36, 39, 185, 221, 224
   fractional 213
Moldflow 162
Monsanto curemeter 214
Mooney plastimeter 213
Moulding powders 48

Netlon 124
Newtonian flow 38, 119
Nomex 203, 221
Non-Newtonian flow 38, 119, 160, 164
Non-Troutonian behaviour 119
Noryl 166
Nylon 12, 119, 123, 166

258 Index

Oil 22–8
  crude 22
  content 134
  distillation 23–8
Orientation in polymers 117, 142, 182, 184, 188, 239
  biaxial orientation 118, 133
Origins of polymers 22
Ozone 57

Paddle mixer 63
Parison, *see* Blow moulding
PEEK 223, 233
Permeability 122, 134, 135, 136
Perspex, *see* Polymethylmethacrylate
Petrol tanks 133
Petrochemistry 22–9
Phenol-formaldehyde 48, 173, 177
Physical properties of polymers 12, 17–22
  time dependency, *see* Creep
Plasticizers 56, 244–5
Plastigel process 252–3
Plastisols 37, 48, 246–52
  coating 248–9
  chemical embossing 251
  curing 249
  foamed 246–7
  formulation 246–7
  mixing 247–8
    see also Mixing
  processing 247–53
  substrates 246
Poly vinylidene chloride 121, 134
Polyacetal, POM 13, 166
Polyacrylonitrile 8
Polybutylene 193
Polybutylene terephthalate, PBT 171
Polycarbonate, PC 11, 13, 51–3, 119, 166
Polychloroprene (Neoprene) 186, 193
Polyethylene 7, 8, 13, 144
  LLDPE 96
  LDPE 15, 119, 133, 166
  HDPE 15, 119, 133, 166
Polyethylene terephthalate, PET 11, 119, 123, 133, 134, 136–7
Polymerization 7–12
  addition 7
  step growth (condensation) 10
  Ziegler–Natta 10, 15
Polymers, nature of 1–29
  advantages 1
  amorphous and crystalline 13, 185
    see also Crystallinity
  atactic 9, 14
  chemical types 6
  dimensions 3
  dispersivity 5
  drawbacks 1
  engineering grades
  frictional properties 90
  glassy 15, 21, 185
  isotactic 9
  manufacture of 6
  molecular weight (molar mass) 4, 186
  morphology 12–17
  naturally occurring 2
  rubbery, *see* Rubbers and Rubbery state
  stereoregular 10, 14, 15
  syndiotactic 9, 14
  unsaturated 8
  volume consumption 1
Polymethyl methacrylate, PMMA 8, 13, 119, 144, 166
Polypropylene 8, 13, 119, 166
  antioxidants in 57
  atactic 10, 117
  extrusion calculation 45
  film 120, 133
  hinge 156–60
  impact modified 9
Polystyrene 8, 13, 119, 166
  high impact 15, 144
Polyurethane 48, 186, 218
Polyvinyl chloride, PVC 8, 144, 166, 176, 186, 218, 242–53
  bottles 134
  dry blending 61
  emulsion grades 242
  K-value 244
  sheeting 188
  slush moulding 237
  stabilizers 245–6
  suspension grades 243

Index

particle size 243
window frames 101, 113
see also Plastisols
Power law fluid 39, 95
Pseudoplasticity 38, 39
Pultrusion, *see* Fibre reinforced plastics

Radlite process 234–6
RAPRA 80
Reaction injection moulding, RIM and RRIM 171–3
Relaxation time 107
Resilience 191
Reverse roll coating 35, 37
Reynolds number 72, 125
Rheology of polymer processing 32–41
Ribbon blender 61
Roller die extruder 210
Rotary presses (textiles) 187
Rotary presses (vulcanizing) 211
Rotational moulding 237–241
   comparison with injection moulding 240
   materials 238
   powders 238
   process 238
   products 238–9
   PVC slush moulding 237
   zero orientation 239
Rotocure 211
Rubber processing 207–12
   cracker mill, *see* Two-roll mill
   mixing 207–9
   see also Mixing
   preforming 209–11
   see also Calendering; Extrusion
   scorch 45, 210
   thick calendered sheet 210
   see also Calendering
Rubbers 49, 179, 191–6
   butadiene, BR 192
   butyl, IIR 193
   chloroprene, CR 193
   ethylene-propylene, EPM 194
   EPDM 8, 194
   latex 195
   natural, NR 186, 191–2, 195–6

Neoprene 193
nitrile, NBR 193
pale crepe 196
production of 194
ribbed smoked sheet, RRS 195
standard Malaysian rubber, SMR 196
styrene-butadiene, SBR 192
vulcanizing, *see* Vulcanization
see also Thermoplastic rubbers
Rubbery state 14, 185–90
Rubber technology 191–219
   see also Rubbers; Vulcanization; Rubber processing; Rubber testing
Rubber testing
   energy to rupture 206
   fractional modulus 215
   hardness 217
   Monsanto curemeter 214
   Mooney scorch time 214
   Mooney plastimeter 213
   product testing 215
   set tests 216
   Shawbury curometer 215–17
   tear strength 217
   tensile testing 216
   variable torque rheometer 215

Sandwich moulding 170–1, 173
SBS 186, 217–8
Scorch *see* Rubber process
Sharkskin 109, 190
   see also Extrusion dies
Shear rate 33
   or processes 34–5
Shear stress 33
   in polymer systems 37
Sheet moulding compound, SMC 179, 227–9
   see also Fibre reinforced plastics
Sintering 237
   sinter coating 239
Solution viscosity 6
Sponge rubber, Sorbo 56
Stabilizers 57, 58
Stereoregular polymers, *see* Polymers
Stiffness 131, 169
Stress decay 18
Structural foam 169–70, 173

260 Index

Synthetic fibres 123
  spinning, spinarets 123
  draw resonance 123

Technopolymers 234
Teflon filter bags 203
Tensar 124
Tension stiffening 119, 128, 143
Tension thinning 119
Terylene 11, 123
Thermal diffusivity 50
Thermal properties of polymers 46–7
Thermoforming 138–45
  air slip 142
  applications 143–4
  drape assistance 141
  'global' pack 144
  large technical mouldings 143
  materials 144
  material stresses 142
  mould, female 141
  mould, male 141
  plug assitance 142
  skin and blister packaging 144
  thin-wall containers 143
  vacuum forming 138–42
Thermoplastic rubbers 49, 185–6, 188, 217–18
  Arnitel 218
  Hytrel 218
  processing 218
  SBS 217–18
Thermoplastics 31–2
Thermosetting polymers 31–2, 176–9, 185
  injection moulding of 173
Transfer moulding 183–4
  cavities 183
  injection moulding, resemblance to 183
  orientation 184
  runners 183
  see also Compression moulding
Trevira 11, 123
Troutonian behaviour 119
Two-roll mill 64, 187
  cracker mill 210

Twin-screw extruders 97–102
  CICO 100
  CICT 99
  conjugated 97
  co-rotating 97, 99
  counter-rotating 97, 98
  CSCO 100
  filling ratio 101
  melt conveying 97–100
  meshing 97
  non-conjugated 97
  non-meshing 97
  output 100

Unsaturated polyester, uPE 48, 174, 221
Urea-formaldehyde 48, 173, 177

Vacuum forming, see Thermoforming
Viscoelasticity 21
Viscosity 32–36
  apparent 39, 41
  capillary viscometry 43
  coefficient, see Consistency index
  materials 35
  melts 40–3
  Poiseuille's equation 43–4
  polymers 35
  relationship to processing 34
  tensile 36, 119
  units 33–4
Viscous dissipation 45–6, 160
Vulcanization 49, 56, 196–201
  accelerators 197, 199
  continuous, CV 212
  efficient, EV 198–201
  formulations 200, 201
  historical 196–8
  marching cure 201
  metal oxides 198
  peroxides 198
  processes 211–12
  reversion 198, 200
  steam curing 212

Z-blade mixer 63, 229